Is There a Universe?

By the same author

Les tendances nouvelles de l'ecclésiologie

The Relevance of Physics

Brain, Mind and Computers
(Lecomte du Nouy Prize, 1970)

The Paradox of Olbers' Paradox

The Milky Way:An Elusive Road for Science

Science and Creation: From Eternal Cycles to an Oscillating Universe

Planets and Planetarians: A History of Theories of the Origin of Planetary Systems

The Road of Science and the Ways to God
(Gifford Lectures: University of Edinburgh, 1975 and 1976)

The Origin of Science and the Science of its Origin
(Fremantle Lectures, Oxford, 1977)

And on This Rock: The Witness of One Land and Two Covenants

Cosmos and Creator

Angels, Apes and Men

Uneasy Genius: The Life and Work of Pierre Duhem

Chesterton: A Seer of Science

The Keys of the Kingdom: A Tool's Witness to Truth

Lord Gifford and His Lectures: A Centenary Retrospect

Chance or Reality and Other Essays

The Physicist as Artist: The Landscapes of Pierre Duhem

The Absolute beneath the Relative and Other Essays

(continued on p. 138)

IS THERE A UNIVERSE?

The Forwood Lectures for 1992

Stanley L. Jaki

LIVERPOOL UNIVERSITY PRESS

Published by
LIVERPOOL UNIVERSITY PRESS
P.O. Box 147, Liverpool L69 3BX

1993

Jaki, Stanley L. (1924-)
Is there a universe?

1. The universe in modern scientific cosmology. 2. Modern philosophies and the universe. 3. Proof of the existence of the universe.

British Library Cataloguing-in-Publication Data
A British Library CIP Record is available
ISBN 0 85323 498 1 Cased
0 85323 009 9 Paper

Printed in the United States of America

Contents

Foreword

The six chapters of this book are the enlarged form of three lectures presented, in November 1992, at the University of Liverpool in the framework of the lectureship established in the will of Sir William B. Forwood (1840-1928), a leading citizen of Liverpool. It is my pleasant duty to express to the members of the University Committee on the Forwood Lectureship my appreciation for the invitation and for the various ways in which they made my stay in Liverpool a rewarding experience.

The terms of the lectureship call for a short course in "non-sectarian divinity." By "sectarian" Sir William Forwood obviously meant attachment to some form of revealed religion. Undoubtedly, all major monotheistic religions claim to be heirs to supernatural revelation, though always with a very natural ground. The ground is that universe whose consideration—so goes a message common to believing Jews, Christians, and Muslims—can lead the mind to the recognition of its Creator. They all hold that because the Creator is so much above Nature as to be able to give reality to it, He can also, if He chooses, impart to mankind a special revelation that goes well beyond what may be gathered about Him and His purposes from the book of Nature.

Theological discourse about such a revelation has always lacked firm basis and consistency in the measure in which attention failed to be given to the reality of the universe. The universe remains also the ground that alone can assure consistency to scientific as well as philosophical discourse. One aim of the lectures was to portray this connection both with respect to science and philosophy. Baffling as this may appear, modern scientific cosmologists failed to ponder the question whether the cosmos they profusely talk about is truly the totality of things or Universe (a word capitalized in this book whenever the context demands it). To anyone familiar with the main trends in modern philosophy, the scant attention, and all too often sceptical scorn, given there to the universe should not seem surprising.

The other main aim of the lectures relates to proofs of the existence of a true totality of material things, or the physical universe. To the proof based on the basic role which universals play in human knowledge, another, perhaps more appealing to the non-metaphysical climate, was added. The convincing character of that proof depends, however, on the espousal of considerations which, though fully rational, differ greatly from the first steps in reasoning as urged by most modern philosophers. But with readiness to take the truly first step about knowing reality, the road will appear wide open to the greatest physical reality which is the Universe. With one's mental eyes riveted on that reality, an infinitely greater and far from "sectarian" Reality may readily emerge on one's mental horizon.

May 13, 1993 S. L. J.

1

A New Science

Is there a universe? The answer may appear straightforward. There has to be a universe if scientists keep writing books with the word "universe" in the title. Public credibility of what scientists say and claim seems to know no limits. Otherwise the editors of *Forbes* would not have paid *The New York Times* large sums for a full-page advertisement in 1991. Much of that page is occupied by a picture of a pig with two wings, followed by the caption in big bold letters: "In a courtroom anything will fly if a scientist testifies to it."[1] Not that the advertisers felt that the credibility of scientists needed any promoting. Their aim was to make the public buy more copies of *Forbes* which happened to carry an exposure of "junk" scientists. The advertisers rightly assumed that the unmasking of such impostors would catch the public eye. Still greater was their confidence in their public-relation strategy riveted in science. The caption quoted above witnessed to their belief that anything said by genuine scientists enjoys built-in credibility all across the board.

One wonders what the public would think if scientists, who had put the word "universe" in the title of their books, started speaking of "universology." With such a word they might emphasize that their subject is really that totality of things which the word "universe" presumably stands for. Such a totality, as will be seen, is most closely tied to respectable science. Yet the use of "universology" might turn into a mixed blessing or something far worse. Suffice to think of that disreputable discourse that has for some time been parading under the name "scientology." Fortunately, dubious adjectives, such as "joyous," are still attached to the word "cosmology" if one is to distract from its, at present, genuinely scientific connotation.[2]

While the word "cosmology" first gained popularity among philosophers, scientists have in recent decades largely expropriated the word to themselves. This is not to suggest that books and articles written by twentieth-century scientists with the words "universe," "cosmos," or "cosmology" in the title, would give the impression that they mean the strict totality of things. This ambiguity was already in full view when around the turn of this century the word "cosmogony" reached its peak of popularity in scientific parlance. By "cosmogony" scientists at that time meant the genesis of the solar system or at most the formation of galaxies. Earlier a resumption by Lord Rosse of Herschel's search of galaxies and, subsequently, the photographing, in very large numbers, of spiral galaxies made fashionable the expression "island universes"[3] that drew attention away from the universe itself.

A telling example of this process was Sir James Jeans' *Astronomy and Cosmogony,* published in 1928. Possibly the last great scientific book with "cosmogony" in its title, it was not about the evolution of the cosmos, but only about the evolution of stars and of galaxies or "other universes." As one who felt at ease with the expression "other universes," Jeans could hardly be expected to have second thoughts for not taking the words "cosmos" or "universe" for the strict totality of things.[4]

It was with the same imprecision that scientists used, during the 17th, 18th, and much of the 19th centuries, the words *mundus, monde,* world, and *Welt.* They invariably failed to offer any of those words as standing for the strict totality of things. Much less did scientists raise the question whether it was possible to know such a totality. The word *mundus* or its equivalents in the titles of well known works by Kepler[5] and Descartes[6] stood for a restriction of the subject matter to the solar system. The few groping paragraphs which those two offered about the stars can easily be understood, writing as they did in studied neglect of the telescope of Galileo. Less forgivable should seem Galileo's restriction of the term *mundus* to the solar system as his telescope showed myriads of stars, especially in the Milky Way.[7] The same restriction was unforgivable in Newton's *Systema mundi,*[8] written as it was more than a hundred years after Galileo's telescopic revelations about the stellar realm. Throughout those hundred years, stars were often referred to as so many other worlds.[9] Most anachronistic was the use of *monde* in Laplace's *Le système du monde,* that went through six editions between 1796 and 1835.[10] While those years largely coincided with Herschel's spotting countless galaxies, Laplace took the solar system for the world.

Less imprecise during the 18th century was the use of the words *coelum,* heaven, and *Himmel.* They usually meant the whole heaven or the entire stellar world as far as it was known. This is very clear in the case of Herschel whose lifework was aimed at finding the very limits of what he called "the construction of the heaven." Although Herschel was ready to "speculate too much rather than too little,"[11] he never asked the question whether it was possible to know that there was a strict totality of galaxies which he took for the basic building blocks of the universe.

Curiously, for many years Herschel thought that he was the first to explain the true reason for the visual appearance of the Milky Way. The first to do so in print was Thomas Wright who meant by the term "universe" the strict totality of all celestial bodies, or galaxies. He did not ask on what grounds he could know that there was such a totality. Worse, he presented all those galaxies as abodes of eternal bliss or punishment.[12]

Immanuel Kant, who learned from Thomas Wright much of what he was to say about the Milky Way, meant the strict totality of things as he used the word *Himmel* (heaven) in the title of his cosmogonical work, *Allgemeine Naturgeschichte und Theorie des Himmels,* published in 1755. Kant remained far removed from a genuinely scientific heaven when in the same work he tried to go Newton one better by "explaining" the evolution of the solar system. Kant roamed across a science-coated fantasy world as he listed in the same work the physical and moral characteristics of the inhabitants of each of the planets.[13] He saw in the universe the strict totality of things not on the basis of science, of which his knowledge was exceedingly meager, but for quasi-theological reasons. These made him assert that the material universe had to be infinite and eternal (once created) because God's infinity demanded an infinite creation. Young Kant certainly provided a telling backdrop for his post-*Critique* image to be taken up shortly.

Johann Heinrich Lambert, the third independent discoverer of the reason for the visual appearance of the Milky Way, mixed fantasy into good science only to the extent of speaking of all future astronomers as traveling from the neighborhood of one star to many other such neighborhoods. But his book, *Cosmologische Briefe über die Einrichtung des Weltbaues* (1761),[14] remained, for a century and a half, the only scientifically consistent (in intent at least) treatment of the totality of gravitationally interacting things. Lambert took the universe, which he called "world edifice," for a finite and hierarchically organized unity of all Milky Ways and of their groupings into more encompassing systems. All those systems were orbiting around their respective central bodies and ultimately around one absolutely central body. Lambert assumed that the orbital speeds were perfectly balanced by the gravitational pull towards the center. Dissipative effects, destructive of such a perfect balance and of the unity of the universe, became considered only a hundred years later.

Lambert's claim about the finiteness of the universe rested on his firm belief in the impossibility of an actually realized infinite number or quantity.[15] While many before Lambert, beginning with Plato and Aristotle, had recognized such an impossibility, it was Lambert who first made use of this in a cosmological context. Still the statement that the world has to be finite insofar as it consists of bodies that can be counted, is merely a first step towards answering, in the affirmative, the question, "Is there a universe?," or a strict totality of things, a question which Lambert never considered.

Lambert had already been dead for three years when Kant startled the intellectual world by his claim in the *Critique of Pure Reason* that the world, or universe, taken for the strict totality of things, was the illegitimate product of the metaphysical cravings of the intellect. He based this claim on the first two of his four antinomies.[16] Curiously, he failed to refer in his discussion of the first antinomy, whether the universe was finite or infinite, to two problems of an infinite homogeneous universe of stars, although those problems had by then been repeatedly discussed and were widely available in print. One problem—originally set forth by Bentley, who consulted Newton and, strangely enough, could not get his agreement —was that in an infinite universe of stars, homogeneously distributed, the gravitational force had to be either infinite or zero. Bentley could offer this consideration only in a qualitative or non-mathematical form.[17] The basis for a convincingly mathematical formulation of the gravitational paradox was provided only in 1828 when Green presented the theory of potentials. The second problem, which later came to be called Olbers' Paradox,[18] had been raised first by Halley before the Royal Society, with Isaac Newton in the chair. The problem consisted in the inference (valid under certain assumptions not questioned for some time yet) that in an infinite homogeneous universe the starlight had to be of infinite intensity at any point and at all time. The very darkness of the night sky could seem therefore to contradict the presumed infinity of the universe.

Kant was either ignorant of these two paradoxes or deliberately ignored them. In either case it remained true that according to the science of the time, though not according to Kant, it was possible to decide the question whether the universe was infinite or not and therefore to have some valid knowledge about the universe. As far as the best science of Kant's time indicated, the universe could not be infinite in terms of the only geometry, Euclidean, known at that time. Therefore, since the first antinomy is not demonstrative (and the second, whether the universe is atomistic or continuous, is irrelevant), Kant's claim that the universe is not a valid conception could not be conclusive either. However, Kant succeeded in instilling in many modern minds a profound doubt whether it is possible to demonstrate that there is a strict totality of things, or universe.

It is indeed by undermining modern man's notion of the all or universe that Kant lived up most to his characterization by Moses Mendelssohn, who three short years after the publication of the *Critique* described its author as the one who "shattered all."[19] When less than a hundred years after Kant's death, his philosophy experienced a revival in the form of Neo-Kantianism, scientists were busy adding scientific veneer to the destructiveness of the first antinomy, without knowing what they were doing. Around 1900 or so it passed for a scientific verity that the Milky Way constituted that part of the universe which alone was observable. Science, so the dictum went, had nothing to do with the infinite stellar realm beyond.[20] This schizophrenic approach to the universe was codified by Lord Kelvin who argued, in terms of mathematical physics, that the infinite "outer" part of the universe could have neither optical nor gravitational influence on the "inner" part, puny though it was by comparison. To make matters more pathetic, Kelvin provided, in the same context, a proof, elegant in its mathematical simplicity as based on Green's theory of potential, that an infinite "Newtonian" universe was subject to the gravitational paradox.[21]

This subtly schizophrenic thinking about the universe was hardly conducive to taking the universe for the subject matter of a scientific discourse keen on consistency. Cosmology as a science, taking the latter for a consistent discourse, could appear an even more hopeless enterprise when around the same time, in the late 1890s, Boltzmann published his two-volume classic on statistical gas theory that came to a close with an application to the science of the universe.[22] In what Boltzmann offered, scientific discourse about the universe lacked coherence in the same way in which the picture he drew about the universe embodied incoherence on a cosmic scale.

Boltzmann spoke of an indefinite number of universes, all separated from one another by a distance 10^{100} times the distance of Sirius which is a mere seven light years.[23] (Today, when the size of the universe is thought of in terms of ten or so billion, or 10^9, light years, the exponential figure given by Boltzmann should seem to prostrate even the boldest imagination). Boltzmann offered no hint as to how many such universes formed the ensemble or the Universe, writ large. Perhaps he took this problem for a philosophical one, convenient to ignore, although it was strictly scientific: Unless the ensemble is definite, statistical considerations about its constituents would hang in mid-air. Such an observation is no less valid today when statistical cosmologies are studied with far greater refinement than Boltzmann would have thought possible.

In Boltzmann's science of the universe there were only universes, all with their separate "languages" or sets of laws. In other words, they all followed their own particular courses of development any of which could be radically different from the one observed in ours. Further, Boltzmann claimed, with an eye on the immense separation among those universes, that nothing could ever be established observationally about those differences. He noticed something of the irony. In a distinctly defensive tone he turned to his reader: "Very well, you may smile at this; but you must admit that the model of the world developed here is at least a possible one, free of inner contradiction, and also a useful one, since it provides us with many viewpoints."[24] But Boltzmann failed to consider how those many viewpoints could be correlated in a single science about an all-inclusive universe, if they indeed reflected very different "languages" or sets of physical laws.

He also overlooked some problems relating to each of those universes. First, he did not explain why within one universe all events did conform to one single direction of time. Second, Boltzmann contradicted his foregoing claim about the unobservability of other times in other universes as he added that his theory "gives an incentive not only to speculation, but also to experiments." Since he was talking about the universe, it was highly inconsistent on his part to list as such possible experiments ones that dealt with "the limit of divisibility, the size of the sphere of action, and the resulting deviations from the equations of hydrodynamics, diffusion and heat conduction."[25] All these experiments could at best refer to one, that is, our universe, but to none of the other universes, let alone to their ensemble whose image, the image of the Universe, emerged only once in Boltzmann's classic as he spoke about the "entire universe, the aggregate of all individual worlds."[26] Possibly, never in the entire scientific literature did the all-encompassing entity receive a treatment as cursory as well as inconsistent as it did in Boltzmann's hands.

Such an entity, if it is to be spotted, calls for philosophical eyes, eyes usually very weak in scientists. Boltzmann's scientific oversight is another matter. He overlooked both aspects of an alternative: If the cosmic aggregate, as spoken of by him, was infinite, the gravitational paradox loomed once more large. If the aggregate was finite, it was bound to dissipate itself into infinite spaces. This latter outcome was pointed out in the very paper in which the science of cosmology had at long last come of age. The paper was Einstein's famed memoir, published in 1917, on the cosmological consequences of general relativity.[27] Only the Universe was still to be accounted for.

For to know that the total number of stars, or the total amount of gravitational matter, cannot be infinite if their distribution is homogeneous in a Euclidean space, is not the same as to know that there is a strict totality of things. The first knowledge is scientific, that is, a knowledge about some quantitative properties of things already existing. The other knowledge, or to know that there is a strict totality of things, is philosophical. Herein lies a principal reason for so many scientists' baffling unconcern with our question. No less a scientist than Einstein recalled with grudging approval the saying that "the man of science is a poor philosopher."[28] The saying was true of him and especially in two respects most relevant to the topic of these chapters.

Einstein did not ask how he knew that there was a Universe, a strict totality of things. Nor did he ask about the relation between that Universe and the universe which is the total mass of gravitationally interacting things as he estimated, on the basis of the average density of matter, the amount of that mass. Yet that Einsteinian universe, as will be seen shortly, was an immense step beyond the idea of an infinite Newtonian universe, as entertained during the 19th century, and can therefore be considered the moment of birth of a new science, a genuinely scientific cosmology.

In this case too, the true birth was prepared by moments indicative of a premature birth. In fact some of what Einstein stated in that famous memoir of his was almost in substance anticipated—and even with some mathematical apparatus—in a remarkable chapter of a book, published in 1871. Unfortunately, the book's general subject was as dubious as its author was scientifically illustrious. The author was none other than the first to occupy a chair of astrophysics anywhere in the world, J. F. K. Zöllner, professor at the University of Leipzig. The book, dedicated to the dubious proposition that the tail of comets represented the basic form of matter, contained a chapter in which Zöllner expanded on a suggestion, first made in 1853 by Bernhard Riemann, to use four-dimensional geometry in physics and even in cosmology. Zöllner showed in some detail that one could thereby obtain a solution to the gravitational paradox which arises in an infinite universe where stars are homogeneously distributed.[29]

That all this could provide some answer to the question, "Is there a universe?," was indirectly noted in 1874 in a public lecture given by W. K. Clifford, professor of mathematical physics at the Imperial College, London. He greeted the first application of four-dimensional geometry to gravitation as an achievement whereby "the universe, as known, becomes again a valid conception."[30] Clifford, it is well to recall, was an early champion of Neo-Kantianism in Britain. Curiously, Clifford failed to see the anti-Kantian thrust of his astute observation. Similarly unperceptive were the many readers of that observation, widely available in Clifford's collected popular essays on science.

Thirty or so years after the publication of Zöllner's book, Einstein had been even more specifically anticipated when K. Schwarzschild, with no reference to Zöllner, took the average density of matter in the Milky Way and evaluated on that basis the curvature of its gravitational field. Schwarzschild sought along that way the resolution of the paradox of an infinite space outside the Milky Way.[31] That Schwarzschild took the Milky Way for the totality of matter became particularly clear from a subsequent publication of his where he spoke of "the universal significance" of this approach to cosmic matter.[32] He was not, however, to be bogged down by strict definitions of totality, let alone by the manner of knowing that such a totality does indeed exist.

Nor did Einstein, as was already noted, allow himself to be encumbered by such considerations when he published his memoir mentioned above. Perhaps Einstein sensed that neither theoretically, nor observationally, can the scientific cosmologist provide assurance that he really deals with the strict totality of things. Still, Einstein's memoir represented an enormous step forward. His theory offered the first contradiction-free scientific discourse about a totality of things subject to the inverse square law of gravitation as manifesting a central field of force. The merit of that achievement was not essentially weakened by the fact that the totality in question was, as late as 1917, readily identified with our Milky Way which appeared to contain all the spiral galaxies, photographed in ever larger numbers.

The universe as accounted for in Einstein's memoir had at least one great immediate advantage over the so-called infinite Newtonian universe. The latter, as was already noted, embodied a serious contradiction, known as the gravitational paradox.[33] Einstein himself took some problems inherent in the "Newtonian universe" as a background to his own achievement. He was wrong in claiming that Newton had set forth those problems. Still it was true that if a three-dimensional distribution of matter was to be a physical possibility, its average density had to decrease at a rate greater than the rate set by the inverse square of distance, and therefore become zero at infinity. Such was Einstein's roundabout way of talking about the gravitational paradox of an infinite universe in which matter is evenly distributed and obeys the inverse square law of gravitation. Einstein should have clearly stated that such a universe, since the gravitational potential is infinite in it at any point, cannot physically exist. Only a year later, in his first popular book on relativity, did Einstein spell this out. He did so by presenting the infinity of that potential in a brief mathematical demonstration which closely resembled the one already offered by Lord Kelvin.[34]

Einstein was no longer alive when documentary evidence had been brought to life about young Newton's conviction that the universe was finite.[35] Einstein (and his distinguished colleagues) could have, however, easily learned that the same conviction was cherished by the mature Newton as well. This fact had been on record through the writings of the two foremost publicists of the 18th century. One of them, J. Addison, was part of Newton's inner circle, in addition to being the editor of *The Spectator*, the most widely read literary circular in the English-speaking world. Addison would not have written in the July 9, 1714, issue that "there is no Question but that the Universe has certain Bounds set to it," had he not felt assured that he echoed Newton's conviction. The "bounded" character or finiteness ascribed to that universe was not weakened by Addison's rhetorical question: "when we consider that it [the universe] is the Work of an infinite Power, prompted by an infinite Goodness, with an infinite Space to exert itself in, how can our Imagination set any Bounds to it?"[36] Voltaire, who adored Newton even more than Addison did, tersely wrote in his *Elémens de la philosophie neutonienne* that "according to Newton and reason the world is finite."[37]

This statement of Voltaire, carried far and wide in the two dozen or so editions of the *Elémens* for the rest of the 18th century, was suddenly forgotten as the 19th century got under way. It was that century that witnessed the curious birth of the idea of an infinite "Newtonian" universe.[38] In one respect, however, Voltaire's pontification about the Newtonian universe was faithfully adhered to even then. Just as Voltaire used Newton's name as a dispensation from proving that the universe was finite, many 19th-century philosophical and scientific gurus found in the slogan, "Newtonian universe," a convenient excuse for not proving the alleged infinity of the universe. They did not even care, for the most part, to call attention to the two grave paradoxes undermining that infinity. By never asking the question, "Is there a universe?," they, of course, took to the shallow waters where Voltaire and many others had already found relief from potentially disturbing philosophical depths.

But to return to Einstein who went on with the somewhat dubious statement that according to Newton the universe "may possess an infinitely great total mass."[39] Einstein was, of course, right in stating that if Boltzmann's statistical gas theory was applied to a finite three-dimensional universe, all matter within that universe would vanish into infinity. This indicated that Einstein's specific concern was to assure coherence to that totality of gravitational matter the sum of which his theory allowed to be calculated. But, as was already noted, Einstein did not raise the question whether there was a difference between that sum and the true total sum of matter, or Universe, writ large. As one who never questioned the existence of the totality of things and did not see in that existence a profoundly philosophical question, Einstein could but take lightly that difference.

Before presenting his own idea about cosmic coherence, Einstein suggested a variant of the Newtonian situation. It amounted to casting matter into a cylindrical space-time manifold. This is what he meant by saying that "the solution corresponds to an infinite extension of the central space, filled uniformly with matter. . . . A universe so constituted would have, with respect to its gravitational field, no center." Consequently, at no point of it could arise an infinitely large gravitational potential even if the gravitational matter was infinite in amount. Nor was the cylindrical universe, in which one would not have to assume a zero density at infinity, in conflict with Boltzmann's statistical mechanics. It was, of course, another question, not considered by Einstein, whether the amount of matter could be really infinite.

Einstein's final remark on that variant of the "Newtonian" universe should be quoted verbatim because the concern expressed in it governed the cosmology he was proposing: "With a definite but extremely small density, matter is in equilibrium, without any internal material forces (pressures) being required to maintain equilibrium." For reasons (and distinctly non-scientific reasons) that will be touched upon later, Einstein greatly preferred a static universe. He introduced his dicta on this variant of the "Newtonian" universe with the remark that it was "not to be taken seriously," that it was but a "foil for what is to follow."[40]

What followed is well known. Assuming an isotropical distribution of matter, Einstein cast it into a four-dimensional manifold or, as he put it, onto a hypersurface. The result was a field equation which contained a supplementary term, the famous constant λ. Its role was to counteract the average velocity of stars, a velocity very small compared with the speed of light. Decades later, Einstein spoke of that constant as the "biggest blunder" of his life.[41] The field equation then yielded the formula for the total mass of the universe which he introduced with the words: "The total mass *M* of the universe, according to our view, is finite."[42] The last remark of his memoir showed his concern for assuring the static character of the universe outlined in this new formalism: "The supplementary term . . . is necessary only for the purpose of making possible a quasi-static distribution of matter, as required by the fact of the small velocities of the stars."[43]

Actually, Einstein's concern was deeper and very philosophical at that, but he failed to articulate it properly for the rest of his life. What followed was foreshadowed in his letter of August 20, 1918, to his friend, M. Besso. There he stated that "since the universe is *unique*, there is no essential difference between considering lambda as a constant which is peculiar to a law of nature or as a constant of integration."[44] What he really meant was that the uniqueness of the universe stood for that ultimate reality where mathematical formalism and physical properties had to have a one-to-one correspondence at the exclusion of any arbitrary assumption. Yet, while this cosmological concern of Einstein was genuinely philosophical, it was not a concern suggestive of the question: "Is there a universe?" Einstein did not even feel the incongruity when he recalled, in his early seventies, that when at the tender age of thirteen, or around 1891, he had read Kant's *Critique*, it made everything clear to him.[45] Einstein never seemed to have wondered what Kant would have thought of one of his favorite remarks, namely, that "the most incomprehensible thing about the universe is that it is comprehensible."[46] He never probed the immense epistemological, and profoundly anti-Kantian, thrust of that pithy remark of his.

Einstein at most kept asserting with an elemental conviction the existence of the universe. Yet the thrust of his words could be more mystical than analytical: "The individual feels the futility of human desires and aims and the sublimity and marvelous order which reveal themselves both in nature and in the world of thought. Individual existence impresses him as a sort of prison and he wants to experience the universe as a single significant whole."[47] The universe as the pantheistic ultimate whole loomed large in his mind as he defended his pantheism by insisting that while it was imperative to recognize the "miracle" of the universe, and thereby distance oneself from "positivists and professional atheists," there was "no legitimate way of getting any further."[48] His was a gut-feeling about the reality of the universe. He failed to consider whether that gut-feeling of his could legitimately support his long-standing eagerness to find out whether God could have created the universe in another form. Neither science, nor philosophy, but a pantheistic theology was the real wellspring of his resolve "to know how God created this world. I am not interested in this or that phenomenon, in the spectrum of this or that element. I want to know His thoughts, the rest are details."[49] Within that perspective, expressive of hubris, it could make no sense to ask the question, "Is there a universe?"

Possibly because of this, Einstein did not, either in 1917 or afterwards, seem to have had a keen realization of the true measure of his scientific achievement in cosmology. He did not single out the cosmological consequences of the General Theory of Relativity as he claimed that Theory to be specifically his own creation as against the Special Theory of Relativity. This is all the more curious, because in that famed memoir of his Einstein introduced his own cosmological model by first conducting "the reader over the road that I have myself travelled, rather a rough and winding road, otherwise I cannot hope that he will take much interest in the result at the end of the journey."[50] Should he not have exclaimed that the scientific sighting of the universe was the end result of that journey?

Einstein's tone remained perfunctory when in later years he took up the subject of the universe. The tone of others was similar as can be seen from the twenty or so essays written mostly by leading physicists in his honor in the late 1940s in a volume that became a classic.[51] The same is true of the remaining contributors, all of them well known philosophers. Not one of them found it important to hail Einstein for making possible the growth of a new and by then powerful branch on the venerable tree of science. In view of this one should approach with a measure of understanding the insensitivity with which prominent scientific cosmologists of our century spoke of their subject-matter as dealing with the All or with the Whole, and in an emphatic sense at that. In doing so they added instructive pages to an already old insensitivity, more philosophical than scientific.

[1] July 16, 1991, p. D20.

[2] See, for instance, *The Joyous Cosmology* (1962; New York: Random House, 1965) by A. W. Watts, which has for its subtitle: *Adventures in the Chemistry of Consciousness* and is "graced" with a foreword by Timothy Leary and Richard Alpert, the former a now largely forgotten champion of experimentation with hallucigenic drugs.

[3] The expression became widely used following the sightings, in the 1840's, of many galaxies by William Parsons, the third Earl of Rosse. See my *The Milky Way: An Elusive Road for Science* (New York: Science History Publications, 1972), pp. 261-62.

[4] J. H. Jeans, *Astronomy and Cosmogony* (Cambridge: University Press, 1928), where not a single reference is made to Einstein's by then ten-year-old relativistic cosmology.

[5] *Harmonice mundi* (1619).

[6] *Le monde ou traité de la lumière,* written between 1629 and 1633, but not published until 1664.

[7] *Dialogo . . . sopre i due massimi systemi del mondo tolemaico e copernicano* (1632).

[8] Prior to writing a small, popular exposition of his celestial physics under the title, *Systema mundi* (1726), translated a year later as *System of the World*, Newton had already given the title, "System of the World," to the Third Book of his *Principia.*

[9] Among the most notable of those works were P. Borel's *Discours nouveau prouvant la pluralité des mondes* (1657) and B. Fontenelle's *Entretiens sur la pluralité des mondes* (1686). In England J. Wilkins offered *The Discovery of a World in the Moone* (1638) and at the end of that century a work of C. Huygens appeared in English under the title, *The Celestial Worlds Discover'd* (1698).

[10] For details, see ch. 5, "The Nebulous Advance," in my *Planets and Planetarians: A History of Theories of the Origin of Planetary Systems* (Edinburgh: Scottish Academic Press, 1978).

[11] He stated this in his "On the Construction of the Heavens," one of his two epoch-making memoirs. *The Scientific Papers of Sir William Herschel*, ed. J. L. D. Dreyer (London: Royal Society, 1912), vol. 1, p. 223.

[12] For these details in Wright's *A New Hypothesis and Theory of the Universe* (1750), see my *The Milky Way*, pp. 191-92.

[13] Kant devoted the entire Part Three of his *Allgemeine Naturgeschichte und Theorie des Himmels* (1755) to an a priori derivation of the physical and moral characteristics of the denizens of each of the planets. On this point and, in general, on Kant's incompetence in Newtonian physics, see the introduction and notes of my translation of that work, *Universal Natural History and Theory of the Heavens* (Edinburgh: Scottish Academic Press, 1981).

[14] On the basis that an actually realized infinite quantity is a logical contradiction Lambert built up a strictly finite universe of hierarchically organized groupings of galaxies. The constituent bodies of each system were held in a stable orbit by the gravitational attraction of a "dark," that is, not star-like, central body or "Regent." For details, see my translation, with introduction and notes, *Cosmological Letters on the Construction of the World-Edifice* (New York: Science History Publications, 1976).

[15] "Immense as the world may appear, it is restricted to finite numbers." See ibid., p. 46.

[16] It has been repeatedly pointed out that in the antinomies Kant shifts ground time and again, that is, takes now the idealist and then the empiricist standpoint.

[17] For details, see my *The Paradox of Olbers' Paradox* (New York: Herder & Herder, 1969), pp. 60-64.

[18] Largely through the discussion of the paradox by the steady-state theorists, none of whom seemed to consult Olbers' paper, although available in English. See my *The Paradox of Olbers' Paradox*, pp. 231-38.

[19] Mendelssohn seems to have referred to the universe as he spoke of his familiarity with the works of various authors and, in particular, of the "Alles zermahlenden Kants." See *Moses Mendelssohn. Schriften zur Philosophie und Aesthetik* (Stuttgart Bad Canstatt: Friedrich Fromman Verlag, 1974), vol. III/1, p. 3.

[20] "With the infinite possibilities beyond, science has no concern," wrote Agnes Clerke in her *The System of the Stars* (1890; 2d ed.; London: Adam & Charles Black, 1905), p. 349. This statement of the leading historian of 19th-century astronomy accurately reflected the scientific consensus. In a less startling phrasing, that consensus was carried far and wide in

Eddington's article "Stars" in the 11th edition of the *Encyclopedia Britannica* (1911, vol. 25, p. 792).

[21] For details and discussion, see my *The Milky Way*, pp. 276-77.

[22] The two volumes of Boltzmann's *Vorlesungen über Gastheorie* were originally published in 1896 and 1898. They are available in a one-volume translation by S. G. Brush, *Lectures on Gas Theory* (Berkeley: University of California Press, 1964).

[23] Ibid., p. 448.

[24] Ibid.

[25] Ibid.

[26] Ibid., p. 447.

[27] English translation, "Cosmological Considerations on the General Theory of Relativity," in *The Principle of Relativity: A Collection of Original Memoirs on the Special and General Theory of Relativity* by H. A. Lorentz, A. Einstein, H. Minkowski and H. Weyl, with notes by A. Sommerfeld, tr. W. Perrett and G. B. Jeffery (London: Methuen, 1923), pp. 177-88.

[28] "Physics and Reality" (1936), in *Out of My Later Years* (New York: Philosophical Library, 1950), p. 58.

[29] J. F. K. Zöllner, *Ueber die Natur der Cometen: Beiträge zur Geschichte and Theorie der Erkenntnis* (Leipzig: Wilhelm Engelmann, 1872). The chapter in question has the title, "On the finiteness of matter in endless space." This book is ignored in J. D. North's *The Measure of the Universe: A History of Modern Cosmology* (Oxford: Clarendon Press, 1965) where (p. 39) another book by Zöllner is mentioned.

[30] Clifford's lecture, "The Philosophy of Pure Sciences," was given at the Royal Institution in March 1873. For its text, see *W. K. Clifford: Lectures and Essays*, ed. L. Stephen and F. Pollock (London: Macmillan, 1901). See especially, vol. 1, p. 387. This point made by Clifford is not recalled by North although in his book, quoted in note 29 above, he has a subsection (pp. 72-81), "A closed universe. W. K. Clifford and others," in the chapter, "Early Relativistic Cosmology."

[31] "Ueber das zulässige Krümmungsmaass des Raumes," *Vierteljahrschrift der Astronomischen Gesellschaft* 35 (1900), pp. 337-47.

[32] See his collection of four addresses, *Über das System der Fixsterne* (Leipzig: B. G. Teubner, 1909), pp. 22 and 42.

[33] Discussed in my essay, "Das Gravitations-Paradoxon des unendlichen Universums," in *Sudhoffs Archiv* 63 (1979), pp. 105-122. Its English version, "The Gravitational Paradox of an Infinite Universe," appeared in my *Cosmos in Transition: Studies in the History of Cosmology* (Tucson, Ar.: Pachart Publishing House, 1990), pp. 189-212.

[34] Its 15th edition appeared in English under the title, *Relativity: The Special and General Theory: A Popular Exposition by Albert Einstein*, authorized translation by R. W. Lawson (New York: Crown Publishers, 1952), see p. 106. For Kelvin's formulae, see his "On Ether and Gravitational Matter through Infinite Space," *Philosophical Magazine* 2 (1901), pp. 174-76.

[35] "De gravitatione et aequipondio fluidorum," in A. R. Hall and M. B. Hall (eds.), *Unpublished Scientific Papers of Isaac Newton* (Cambridge: University Press, 1962), pp. 89-156. See especially, pp. 138 and 142-43.

[36] *The Spectator.* A new edition, with an introduction, notes, and index by H. Morley (London: George Routledge and Sons, 1891), vol. 3, pp. 472-73.

[37] First published in 1738. It went through two dozen editions during the rest of the eighteenth century. For quotation, see Première Partie. chapitre I, in *Oeuvres complètes de Voltaire* (Paris: Garnier Frères, 1877-82), p. 457.

[38] An early example was Olbers' famous paper on the transparency of the heavens, published in 1823. For text and details, see my *The Paradox of Olbers' Paradox* (New York: Herder & Herder, 1969), pp. 131-43 and Appendix III.

[39] Einstein, "Cosmological Considerations . . . ," p. 178. Einstein perhaps thought of Newton's letters to Bentley. But even there Newton was more eager to disprove Bentley's claim that it was possible to show that the universe was finite. Einstein did not seem to be struck by the fact that in the *Principia* Newton does not claim the universe to be infinite. In fact, Newton says there very little about the universe as such.

[40] Ibid., p. 179.

[41] In a conversation, around 1947, with G. Gamow, as recalled by the latter in his *My World Line* (New York: The Viking Press, 1970), p. 44.

[42] Einstein, "Cosmological Considerations . . . ," p. 187.

[43] Ibid., p. 188.

[44] A. Einstein and M. Besso, *Correspondance 1903 - 1955,* ed. P. Speziali (Paris: Hermann, 1972), p. 134.

[45] As reported by Einstein's boyhood friend, M. Talmey, in his *Relativity Theory Simplified and the Formative Period of Its Invention* (New York: Falcon Press, 1932), p. 164.

[46] Such words are exactly ascribed to Einstein, for instance, by B. Hoffmann, *Albert Einstein: Creator and Rebel* (New York: The Viking Press, 1972), p. 18, and by Y. Elkana, "The Myth of Simplicity," in G. Holton and Y. Elkana (eds.), *Albert Einstein. Historical and Cultural*

Perspectives. The Centennial Symposium in Jerusalem (Princeton: Princeton University Press, 1982), p. 242. Elkana refers to Einstein's lecture, "Physics and Reality" (1936) at the Franklin Institute, but there Einstein stated: "The world of our sense experiences is comprehensible. The fact that it is comprehensible is a miracle." See *Ideas and Opinions by Albert Einstein* (New York: Crown Publishers, 1954), p. 292.

[47] "Religion and Science" (1930), in *Ideas and Opinions by Albert Einstein* p. 38.

[48] A. Einstein, *Lettres à Maurice Solovine* (Paris: Gauthier-Villars, 1956), pp. 102-03.

[49] Several of such utterances of Einstein are quoted by Elkana in his "The Myth of Simplicity," pp. 238-40.

[50] Einstein, "Cosmological Considerations . . . ," p. 180.

[51] *Albert Einstein. Philosopher-Scientist*, ed. P. A. Schilpp (1949; New York: Harper Torchbooks, 1959).

2

An Old Insensitivity

About references by scientific cosmologists to the "all" or the "whole" as their subject matter, the first thing to note is that they were slow in coming. For a dozen or so years after the publication of Einstein's famed memoir in 1917, reactions to it had a note not in tune with its basic thrust. Such was the chagrin which E. Borel voiced in 1921 over the fact that the universe could no longer be considered infinite,[1] a reaction more suggestive of his commitment to some form of pantheism than to serious philosophical stock-taking. Scientific cosmologists even failed to note that scientific cosmology is a branch of science, very different from other branches. Unlike in all other branches, where the object is outside the observer, the scientific observer of the cosmos always remains inside it.

Worse, authors of the first major commentaries on general relativity were wont to sublimate physical reality into world-lines and other abstractions. From there it was but a short and very logical step to speak of the universe as the product of the mind and have thereby the universe disappear in the slippery world of subjectivism if not plain solipsism. Almost symbolically, this came about in the wake of World War I which itself had dealt a devastating blow to a world of culture that still claimed as its own some remnants of the objectivity of earlier centuries. The first and most widely read high-level popularization of general relativity, Eddington's *Space, Time and Gravitation,* first published in 1920, came to a close on a markedly solipsistic note: "We have found a strange foot-print on the shores of the unknown. We have devised profound theories, one after another, to account for its origin. At last, we have succeeded in reconstructing the creature that made the foot-print. And Lo! it is our own."[2]

Any judicious reader could have, of course, objected that since the footprints were pure mental constructions the shores into which they were allegedly imprinted could hardly be made of a more tangible construct, such as real matter. In fact, Eddington had just made a programmatic declaration that left him fully open to that objection which was destructive of his reasoning while it safeguarded material reality and therefore the physical universe. This is not to suggest that Descartes, who tried to derive the physical universe from mere extension, had the right to claim: "Give me matter and motion and I will construct the universe." In quoting these famous words often ascribed to Descartes, Eddington remained unaware that he was merely drawing the ultimate solipsistic consequences of Descartes' rationalism as he stated in a way of contrast: "The mind reverses this. 'Give me a world—a world in which there are relations—and I will construct matter and motion'."[3] Only the shadows of matter and motion were thereby conjured up, with the universe relegated to a conceptual limbo.

The material universe had at most a shadowy existence in the strictly scientific report which Eddington submitted to the Physical Society of London a year earlier. There he took exception to Einstein's world model in which matter produced space, that is, a network of permissible paths of motion. As Eddington correctly noted, in Einstein's "theory, if all matter were abolished, *R* [the radius of the path with the least curvature] would become zero and the world would vanish to a point." Eddington had only gentle sarcasm for that very realist concern: "There is something rather fascinating in a theory of space by which, the more matter there is, the more room is provided." The basic fault of that theory, Eddington continued, was not scientific but philosophical, namely, that Einstein was "unwilling to admit that a thinkable space without matter could exist."[4] The fault lay in fact with Eddington's solipsistic thinking in which there could be no difference between the existence of a mere idea in the mind and some sort of physical existence corresponding to it.

Not that Eddington found himself alone with this cosmic view in which there was no room for a real cosmos or universe. He invoked as star witness Willem de Sitter, the first to make a major contribution to Einstein's memoir on the cosmology of general relativity: "For our part, we feel equally unwilling to assent to the introduction of vast quantities of world-matter, which (to quote de Sitter) 'fulfills no other purpose than to enable us to suppose it not to exist'."[5]

That Eddington chose de Sitter as a witness was appropriate for another reason as well. In most major presentations of general relativity to be published in the early 1920s, the idealist philosophy Eddington held dear merely lurked between the lines. Further, those lines were relatively few. Discourse about the cosmological consequences of general relativity was short in the well known works of W. Pauli,[6] M. Born,[7] and A. N. Whitehead. The latter's book on relativity[8] had no chapter, not even a paragraph, on the universe. In that respect it fully anticipated the thrust of Whitehead's philosophical cosmology, *Process and Reality*, where, as will be discussed later, the universe became amorphous through the claim that it was bound to assume all possible forms through infinite ages. Whitehead's was a dubious strategy which Darwin and the Darwinists made respectable: Profuse talk about the evolution of entities served as an excuse for not facing up to the philosophical issues involved in the reality of those entities themselves. Whatever the pragmatic success of such a strategy, it revealed a philosophical insensitivity all the greater as it related to that greatest and most encompassing reality which is the universe. How could the evolution of the universe be the subject of rational discourse if the question of its existence was swept under the rug?

The cosmologist who did worse than slight that question was de Sitter. By the early 1930s Eddington could have found in de Sitter a spokesman of an idealist construction of the universe bordering on solipsism. In his *Kosmos*, a set of six lectures given in Boston in 1931, de Sitter spoke of scientific laws as the only true reality for the scientist. Furthermore, he characterized as a mere faith, though indispensable for science, the conviction about the unity of all things which he took for a mere hypothesis: "After all the 'universe' is an hypothesis, like the atom, and must be allowed the freedom to have properties and to do things which would be contradictory and impossible for a finite material structure. What we observe are the stars and nebulae constituting our 'neighborhood.' All that goes beyond that, in time or in space, or both, is pure extrapolation. . . . It is not inconceivable that this hypothesis [the universe] may at some future stage of the development of science have to be given up, or modified, or at least differently interpreted."[9]

Such a conceptual volatilization of the universe could not be offset by remarks which, insistent as they were that the new cosmology dealt with the "all," had no explicitly realist ring to them. One such remark occurred in L. Silberstein's survey of early "attempts at a determination of the curvature radius of space-time," published in 1930. In speaking of the symmetry properties of the universe, Silberstein warned that they had nothing to do with aesthetics. His book was meant to be "on Cosmology, or the actual Universe as a whole."[10]

Special attention is deserved by similar remarks, made shortly afterwards by H. P. Robertson, professor of mathematical physics at Princeton University. The printing of his lecture, "The Expanding Universe," in the September 9, 1932, issue of *Science*,[11] carried world-wide his repeated insistence that Einstein's general relativity "deals with the structure of the universe as a whole." Robertson also stressed that his lecture's interest lay in "the universe as a whole and not in the phenomena which occur within the borders of such a closely-knit system as our galaxy." He further emphasized that the red shift of galactic spectra made sense only if one accepted assumptions such as that "all matter in our universe has always maintained a physical unity, even in the remote past—that the actual universe is not a fortuitous superposition of two or more incoherent parts."[12]

Similar was Robertson's insistence on the universe as the "whole" in the introduction of his famed essay, "Relativistic Cosmology," in the January 1933 issue of *Reviews of Modern Physics,* which served as the first "textbook" for most physicists on the subject. The essay, of almost thirty quarto pages, contained a survey of "attempts which have been made during the past decade and a half to solve the general problem of the structure of the universe *as a whole*," he himself adding the emphasis. That to deal with "the whole" was more than a scientific problem, he merely suggested. He did so by admitting that the three major observational consequences of general relativity (the advance of the perihelion of Mercury, the gravitational red shift, and the bending of light) did not give sufficient assurance about the competence of general relativity to deal with the universe as a whole. In fact, he acknowledged that even the recent discovery of the recession of galaxies was not enough to deal with that whole without "such hypotheses as appeal to the general or philosophical predilections of the investigator."[13] About those hypotheses Robertson offered no details, except perhaps a remark towards the end of his article. There he spoke sympathetically of Weyl's a priori postulate, aimed at securing simultaneity in the universe as a whole. Robertson gave no further hint about the manner, obviously non-scientific, in which a scientific cosmologist could assure himself that there was a "whole" or Universe.

If Robertson's reason for not doing so had been his suspicion that his colleagues would not be interested, he would have been on the right track. He could not be unaware of the article, "Models of the Physical Universe," which R. C. Tolman had published in the April 8, 1932, issue of *Science*.[14] The article was the text of a lecture which Tolman delivered at the invitation of the Philosophical Union of the University of California at Los Angeles, a group presumably eager to learn about the philosophical import of the newly born scientific cosmology. That this new branch of science required a departure from a rigid empiricism, that its bold generalizations could be the source of deep intellectual satisfaction, was true enough, but not sufficiently philosophical. Nor did Tolman steer interest toward serious philosophizing as he recalled, at the outset, the facetious, if not malicious remark that "philosophy is the systematic misuse of a terminology specifically invented for this purpose."[15] Whether Tolman's lecture, full of technicalities, could enlighten the Philosophical Union and philosophers at large should seem most doubtful. They were certainly not asked by Tolman to ponder his philosophically most important claim that "for the scientist the problem of cosmology lies in an attempt to describe and understand the structure and behavior of the physical universe *as a whole*"[16] (italics added).

Nor was philosophy directly in view when two years later Tolman published his *Relativity, Thermodynamics and Cosmology* which became, for decades, the most important advanced textbook on the subject. Not that Tolman had not brought his book to a close with a momentous warning: "It is appropriate to approach the problems of cosmology with feelings of respect for their importance, of awe for their vastness, and of exultation for the temerity of the human mind in attempting to solve them." Still, Tolman's concern that those problems, insofar as they are scientific, "be treated by the detailed, critical, and dispassionate methods of the scientist,"[17] was not even a pointer toward our question. There would have been no ground for the concern voiced by Tolman, if scientific cosmologists had been taking the trouble of proving that there is a totality of things that exists independently of their speculating about it.

Among those who spearheaded in the 1920s and 1930s the observational sector of scientific cosmology, philosophical insensitivity to the cosmos could readily turn into a sense of dejection. A gripping illustration of this is the conclusion of *The Realm of the Nebulae*, the text of Edwin Hubble's Silliman Lectures delivered at Yale in the Fall of 1935. That Hubble was unable to take the systematic red-shift of nebulae for an evidence of the expansion of the universe suggests some strange prejudices in his cosmological and philosophical thinking. He deeply distrusted theoretical cosmology. In *The Realm of the Nebulae* there were only two brief references to Einstein and in neither case did he appear as the creator of the new science of cosmology. Pantheistic proclivities may have prompted Hubble's fondness for the idea of an essentially unchangeable, that is, non-expanding universe.

He was so serious in searching for a physical cause, other than recessional velocity, of the galactic red-shift as to characterize "the expanding universe of general relativity" to "persist in theory" as opposed to reality. About cosmic reality he struck a note of strong scepticism: "The exploration of space ends on a note of uncertainty. And necessarily so. We are, by definition, in the very center of the observable region. We know our immediate neighborhood rather intimately. With increasing distance, our knowledge fades, and fades rapidly. Eventually, we reach the dim boundary—the utmost limits of our telescopes. There, we measure shadows, and we search among ghostly errors of measurement for landmarks that are scarcely more substantial. The search will continue. Not until the empirical resources are exhausted, need we pass on to the dreamy realms of speculation."[18]

The insensitivity of leading scientific cosmologists to the crucial philosophical problem embedded in their very subject will appear less startling if one recalls the baffling attitude of some theologically trained colleagues of theirs. One of them, E. W. Barnes, eventually rose from tutorship in mathematical physics in Cambridge to the Anglican See of Birmingham. He would not have earned himself the reputation of being the *enfant terrible* of his Church, had his rationalism been tempered by some good metaphysics. In the cosmology of general relativity he merely saw a liberation from Euclidean infinity and the opportunity which a finite universe offered for fully fathoming it.[19] Although Barnes spoke, in connection with the red-shift of the galactic spectra, of the motion of "the Universe as a whole,"[20] he did not hear the philosophical sound of the chord he had struck thereby. His failure should seem all the more notable as he quoted, almost in the same breath, Clifford's popular exposition of Riemannian cosmology in 1873.

Another theologian-cosmologist was the Belgian priest, Georges Lemaître. In 1927 Lemaître became the first to show that the red-shift of galactic spectra has its natural explanation in the expansion of the universe, a feat for which he received emphatic credit in Robertson's popular lecture, discussed above. A very devout and orthodox Catholic priest with good formation in Thomism, Lemaître capped his illustrious career with the Presidency of the Pontifical Academy of Sciences. He, however, never broached, in his various essays dealing with faith and science, the question of the universe as posed by modern scientific cosmology.[21] The topic escaped the attention also of E. T. Whittaker, a fact all the more regrettable as his writings on science and religion[22] were much relied upon in the address which Pope Pius XII delivered on November 21, 1951, on science and faith.[23]

A dozen or so years later, there appeared J. D. North's history of modern cosmology, under the title, *The Measure of the Universe*. Its author's commodism, mixed with idealism, did not do major injustice to his recounting of the views and speculations of modern scientific cosmologists. In commenting on the phrase, often occurring in the literature, that "true cosmology deals with the All, and nothing less," North took the view that all sciences have a restricted subject matter and that scientific cosmology was no exception to that. He drew, therefore, the conclusion that "for scientific cosmology it is . . . certainly misleading to refer to its subject matter as the universe." There was much truth in this remark which aimed at the difference between the character, however comprehensive, of a given cosmological model and the comprehensiveness which the term, Universe, is meant to convey. And it was at least partially true that "at the very least it should be generally acknowledged that the full significance of the term 'universe' cannot be given in terms of denotation alone."[24] But the partial truth of this could only be threatened if the definitional use of the term was meant to discredit any talk about the reality of absolute physical totality.

Obviously, it was one thing to speak of an absolute totality of things and another to demonstrate that there was such a totality. Still, one wonders whether it was not by intent that North, who quoted a book of Zöllner on gravitation as well as a remark of Clifford on Riemannian cosmology,[25] had ignored Clifford's far more important remark about the re-validation of the notion of the universe. North may have sensed that his commodist idealism could not be reconciled with Clifford's remark that a crucial philosophical bearing lay hidden in Zöllner's idea on a contradiction-free treatment of the totality of gravitationally interacting bodies in a central field of force obeying the inverse square law.

The year of the publication of North's book, 1965, became, of course, a landmark year in the history of modern scientific cosmology through the discovery of the 2.7°K cosmic background radiation. As a result, scientific cosmology turned into an explosive field of study, but the gap between the observationally explored universe and the universe as true totality failed to be closed even in philosophical intent. Two generations after Hubble, and with a far more accurate knowledge of the astronomical universe than the one available to him, observational cosmologists can still strike a note of doubt echoing his misgivings. "Mapping the universe," wrote Margaret J. Geller a few years ago, "will undoubtedly keep us busy, awed, fascinated for a long time to come. I often ask myself what we will learn about the large-scale structure during my lifetime. There will be surprises, answers to old questions, and the uncovering of new puzzles. At every stage we will think we understand, but at every stage there will be nagging doubts in the minds of those who wonder."[26]

As to the latest crop of theoreticians, they kept their profession's disinterest in a philosophical justification of their talk about the universe. A major illustration of this is *Origins*, a recent series of interviews with twenty-seven noted cosmologists,[27] a book as revealing of the interviewers as of those whom they interviewed. Part of the list of questions was posed to all those interviewed. For instance, they all were asked whether they thought there was any purpose in the cosmos or in existence in general. Only once, in all appearance, was the question raised about the ground that assures the reasonability of scientific discourse about that whole which is the universe.

This is not to suggest that there was much depth (or cogency) in the question, "Do you think that when we use words like 'universe', we understand what those words mean?," a question posed to A. Lindé. He in turn generated little light with his reply: "Different people use this word differently. For anyone who tries to understand it, at least, the universe is everything which exists. However, this may not be the last word in the interpretation of this subject, since, first of all, the question arises whether we should include consciousness in the definition of everything that exists."[28]

Most cosmologists, and especially the ones emphasizing the observational and experimental aspects of their research, would not go even as far as to face up to the question whether there is a universe. This attitude of theirs should seem particularly jarring when seen, for instance, against the opening chapter, "The Universe Observed," in *The Early Universe*, a major survey of cosmological research as it stood at the end of the 1980s.[29] Its authors failed to see any philosophical problem on hand as they discussed quantum cosmology, in spite of the question they interjected: "When was the last time when you saw a Universe created or destroyed?"[30] Nor did they exploit the philosophical significance of their observation that "an equation for the wave function of the Universe no more resolves the issue of the quantum evolution of the Universe, than the Schrödinger equation for an electron resolves the issue of the quantum evolution of said electron."[31]

No wonder. A basic grasp of the philosophical issue on hand would have demanded a sensitivity to the difference between realism and idealism. One would in vain look to scientific cosmologists for an unambiguously realist pondering of the question whether there is a universe. A major evidence of this is on hand in P. Kerszberg's discussion of "The Einstein-de Sitter Controversy (1916-17) and the Rise of Relativistic Cosmology," the subtitle of his book, *The Invented Universe.* Its concluding pages deal with the difference between "the philosopher's universe and the physicist's." There he noted that all the philosophy embedded in modern scientific cosmology reflects a trend to prefer a universe which is "becoming" to a universe which "is" and expressed sympathy with this trend.[32] The trend is known as idealism, mixed with process philosophy. But to yield to that trend is to lose hold on the "everything" which is the universe by trying to account for everything that takes place in it.

Had he been free of this trend, Hermann Weyl could have offered, around the mid-point of the century, a not-too-late antidote to it. I mean his categorical assertion that "the objective world simply *is*, it does not *happen*."[33] What he tried to say was that the universe was not contingent in the sense in which everything is contingent that happens, the chief evidence of ontological contingency. Once more, as so often in the past, the assertion that the universe was non-contingent and therefore necessary, nipped in the bud interest in proving in the first place that there indeed was a universe. This necessitarian aprioristic approach to the universe inspired a thinking about the universe in which there was no room for wonderment. It became true of the universe as well of what is true of all objects that can be had at will, namely, a thorough loss of respect for it. The process brings unintended support to an utterance of Chesterton who in 1905 summed up the delusion of his times with the thumping phrase: "Everything matters except [the] everything."[34] Most regrettably, the delusion has now taken on that glittering garb which scientific cosmology alone can provide.

No dent is made on that delusion when on a rare occasion cosmologists care to define what they mean by the terms, universe and cosmos, and say, for instance, that the universe is an entity which, "by definition, includes everything."[35] They even ignore the fact that a definition is never a proof of the existence of the entity so defined. Something even more sinister raises its head in the book written by a scientific cosmologist who alone showed sufficient awareness of our question. To begin with, the question, "Is there a universe?," is about its existence and not merely about what the word "universe" properly means. It is shockingly superficial to say, as W. D. Sciama did in his *The Unity of the Universe*, that "since the universe is, by definition, the totality of things, its uniqueness is self-evident."[36] Such a reply will settle matters only for unrepentent Platonists. They refuse to recall the many pitfalls which the history of philosophy, this great testing ground of philosophical systems, has uncovered about Plato's message.

Platonism immediately reveals its dubious character when one infers on its ground the non-existence of other universes. This, of course, is not a concession Sciama was willing to make, because he excluded other universes on precisely such ground: "The laws of nature *must* be formulated in such a way that they relate only to the actual universe, for other universes, by definition, cannot exist"[37] (italics added). Clearly, Sciama tried to achieve far more than to prove that there was a universe. He wanted to demonstrate the existence of the universe by proving in the same breath that the universe must necessarily be what it is. In this way he burdened a not at all easy philosophical task with a heavy dose of a priorism. Worse, he thought that the answer could be given in terms of science. Obviously, as will be seen later, the result was not so much a scientific cosmology as a set of scientifically undemonstrable postulates that were outright pseudo-metaphysical if not something worse. In the process, thinking about the physical universe became loaded with paradoxes far more debilitating than the optical and gravitational ones. In fact, the universe itself became spoken of as a heap of fragmentary units that could carry the label "universe" only through honoring logic in the breach.

[1] In his *Space and Time* (London: Blackie and Son, 1926), an exposition of Einstein's theory of relativity, Borel conceded the finiteness of the universe only as a "convenience."

[2] A. S. Eddington, *Space, Time and Gravitation* (Cambridge: University Press). Reference is to its Harper Torchbook edition, first published in 1959.

[3] Ibid., p. 198.

[4] A. S. Eddington, *Report on the Relativity Theory of Gravitation* (London: Fleetway Press, 1920), p. 90.

[5] Ibid., pp. 90-91.

[6] W. Pauli's presentation of general relativity in the 5th volume of the *Enzyklopädie der mathematischen Wissenschaften* (1921) came out in English translation by G. Field, with supplementary notes by the author, in 1958, under the title, *Theory of Relativity* (New York: Pergamon Press).

[7] This work of M. Born was quickly brought out in English, *Einstein's Theory of Relativity* (London: Methuen, 1922), in the translation of H. L. Brose. In the preface to the original German, Born wholly misconstrued the epistemology underlying general relativity by claiming it to be corresponding to the one set forth by M. Schlick, who, it is well to recall, was the founder of logical positivism which has no room for cosmology.

[8] A. N. Whitehead, *The Principle of Relativity with Applications to Physical Science* (Cambridge: University Press, 1922). The almost total absence in this book of references to the cosmology of general relativity foreshadows the baffling disregard which Whitehead was to show, in terms of his evolutionary idealism, toward the universe in his philosophical cosmology, *Process and Reality*.

[9] W. De Sitter, *Kosmos* (Cambridge, Mass.: Harvard University Press, 1932), pp. 133-34.

[10] L. Silberstein, *The Size of the Universe* (Oxford: University Press, 1930), p. 52.

[11] H. P. Robertson, "The Expanding Universe," *Science* 76 (1932), pp. 221-26. The lecture was delivered before the ninth annual meeting of the West Virginia Academy of Science in Athens, West Virginia, April 29, 1932.

[12] Ibid., pp. 221-23.

[13] H. P. Robertson, "Relativistic Cosmology," *Reviews of Modern Physics* 5 (1933), pp. 62-90. For quotations, see pp. 62-65.

[14] R. C. Tolman, "Models of the Physical Universe," *Science* 75 (1932), pp. 367-73.

[15] He credited Prof. F. Zwicky, who had just begun his career at Mount Wilson Observatory, with giving him the phrase in the German original, "Philosophie ist der systematische Misbrauch einer eigens zu diesem Zwecke erfundenen Terminologie."

[16] Ibid., p. 368.

[17] R. C. Tolman, *Relativity, Thermodynamics and Cosmology* (Oxford: Clarendon Press, 1934), p. 488.

[18] E. Hubble, *The Realm of the Nebulae* (1936; New York: Dover, 1958), pp. 201-202.

[19] See his Gifford Lectures, *Scientific Theory and Religion: The World Described by Science and its Spiritual Interpretation* (Cambridge: University Press, 1933), pp. 49 and 392.

[20] Ibid., p. 380.

[21] See his *L'hypothèse de l'atome primitif: Essai de cosmogonie* (Neuchatel: Editions du Griffon, 1946), a collection of such essays.

[22] See especially his Donnellan Lectures, *Space and Spirit: Theories of the Universe and Arguments for the Existence of God* (Hinsdale, Ill.: Henry Regnery, 1948).

[23] The address, "The Proofs of the Existence of God in the Light of Modern Natural Science," was made widely available in brochure forms

of which one was issued by the National Catholic Welfare Conference (Washington D. C. 1951, 18pp).

[24] J. D. North, *The Measure of the Universe: A History of Modern Scientific Cosmology* (Oxford: Clarendon Press, 1965), pp. 274-75.

[25] Ibid., p. 39. The book was not the one on the tail of comets and Clifford's remark as quoted by North had nothing to do with the gist of that famed chapter of Clifford's book, discussed in the preceding chapter.

[26] M. J. Geller, "Mapping the Universe: Slices and Bubbles," in J. Cornell (ed.), *Bubbles, Voids, and Bumps in Time: The New Cosmology* (Cambridge: University Press, 1989), p. 54.

[27] A. Lightman and R. Brawer, *Origins: The Lives and Worlds of Modern Cosmologists* (Cambridge, Mass.: Harvard University Press, 1989).

[28] Ibid., p. 393.

[29] E. W. Kolb and M. S. Turner, *The Early Universe* (Redwood City, CA.: Addison-Wesley, 1989). The book is volume 69 in Frontiers of Physics Lecture Series.

[30] Ibid., p. 449.

[31] Ibid., p. 457.

[32] P. Kerszberg, *The Invented Universe* (Oxford: Clarendon Press, 1989), p. 393.

[33] H. Weyl, *The Philosophy of Mathematics and Natural Science* (1949; New York: Atheneum, 1963), p. 219. In the same breath, Weyl asserted that all change in the world is a function of the conscious mind!

[34] G. K. Chesterton, *Heretics* (London: John Lane, 1905), p. 13.

[35] J. Narlikar, *The Structure of the Universe* (Oxford: University Press, 1978), p. 1.

[36] W. D. Sciama, *The Unity of the Universe* (1959; Garden City, N. Y.: Doubleday, 1961), p. 161.

[37] Ibid., p. 204.

3

Universes and Universe

The kind of universe which, according to Sciama, had to exist necessarily was, of course, the one embodied in the idea of a steady-state universe. Its chief supporters quickly succeeded in gaining for it wide renown which, in an increasingly publicity-conscious age, readily passed for intrinsic merit. Many were swayed, though not Einstein, the creator of the new science of cosmology. In one of his rare utterances on the steady-state theory he dismissed it as a "romantic speculation."[1] He should have rather characterized it as an abuse of plain logic. Beyond that abuse there lay that supreme ignominy, hardly ever noted, which the universe could conceivably suffer, namely, its breaking into incoherent parts.

The very notion of the universe fared very badly in the hands of steady-state theorists. Not that this could be seen firmly outlined in the distinction by which H. Bondi, a chief spokesman of the steady-state theory, brought to a close the Introduction of his *Cosmology*. Bondi began by distinguishing between two totalities. One was the totality of physical events "which may affect us at some time in the future and all events which have been or will be affected by us." What he had in mind was the ever widening range of man's observations of cosmic reality, or that totality of things or "universe" which is actually or potentially observed. To distinguish this totality or "universe" from the absolute totality of things or "Universe" was certainly valuable. But the value of this distinction was undercut when in the same breath Bondi turned to "the largest set of all physically significant objects" and noted that it "may include points that have never been and will never be physically linked to us and may exclude points which are so linked."[2] Bondi failed to realize that it was meaningless to speak of something that was unknowable by definition. He thereby came, in more than one sense, under the sway of Wittgenstein's devastating dictum: "Whereof one cannot speak, thereof one must be silent."[3]

Just as the steady-state theorists did not think of what was more obvious, from the logical point of view, in Wittgenstein's dictum, they also overlooked that they gave a physical application to it. They did so by dividing the universe into two parts. One finite and observable, the other infinite and unobservable by definition. As physicists they should have recoiled from attributing eternal sameness to the Universe as the chief aim of their cosmology. Eternity is not a subject matter of a physics which takes observational evidence for its supreme touchstone of truth. Once more, the postulating of physical eternity demanded an impossible price. It should be enough to think of the claim of steady-state theorists that hydrogen atoms were popping, at a steady rate, into existence out of nothing. With this claim of theirs they tried to offset the decrease of the average density of matter in an expanding universe. What they really wanted to establish by speaking of the "perfect" cosmological principle, was the sameness, on a large scale, of the universe at all times. They would have promoted clarity if they had identified themselves as dogmatic eternalists. As will be seen, these have always been short on proofs but very prolific with their *fiats* about the alleged eternity of the universe.

If not science, which can never prove eternity, clarity at least was served by Bondi, who insisted that the emergence of hydrogen atoms was not a transformation of energy into matter, but a coming into existence of matter literally out of nothing and without the action of a Creator.[4] But this farcical mechanism, which flew in the face of theology as well as physics, burdened the steady-state physicists with the task of disposing of an infinite amount of matter piling up since eternity. They did so by falling back on a pseudo-divine *fiat.* Galaxies, they claimed, constantly moved beyond the horizon set by the line where the recessional velocity became infinite, and disappeared in a cosmic sink. An unusual sink it was, as it had to contain a multiply infinite amount of matter. For not only the amount of hydrogen was to grow infinite through eternity but also the matter which passed through the horizon where its relativistic mass became infinite.

The steady-state theorists, who boldly spoke of the emergence of infinite matter through infinite times, and out of nothing, in the observable part of the universe, did not become so bold as to pump that matter out of that bottomless sink. Much less did they muster courage to claim that the sink was a mechanism of annihilation. At any rate, the universe piling up into infinities within that sink was by definition unobservable. Such was a high-handed splitting of the universe into two parts, one finite and observable, the other unobservable though infinite. Compared with this the late-19th-century partition of the universe into a puny central part and an infinite and practically unobservable part surrounding it, should seem an innocent vagary of the scientific mind.

The steady-state theorists made no effort to make use of the Copenhagen interpretation of quantum mechanics, to justify "scientifically" the creation out of nothing of hydrogen atoms at a steady rate through eternity. The first to make that principle the cornerstone of their cosmology were the proponents of the multiworld theory. Undoubtedly, the act of observation, insofar as it implies the impingement of light rays on the object to be observed, becomes part of the measurement and of its "uncertainty" or imprecision. But this is not the same as to state that the consciousness of the observer influences the measurement, let alone that it creates the reality through its being observed.

So much about the very roots of the highly technical multiworld cosmology.[5] The latter implies a fragmentation of the universe not so much into countless universes as into as many universes as there are observers. The abysmal depths of subjectivism into which the universe is thereby made to sink are also very disconnected depths. They can be glimpsed by looking at a book with the phrase "other worlds" in its title. Its author couples a praiseworthy declaration of the totality represented by the universe with a question which makes it a function of the thought of a mere individual: "When we are dealing with the whole universe—all of creation—there is, by definition, nothing external that can observe it. The universe is supposed to be everything that there is, and if all is quantized, including space-time, what can collapse the cosmos into reality without invoking consciousness?"[6] From there it is but one step to abusing the science of cosmology by presenting it in a book with the title, *Conscious Universe*.[7] Beyond that the sole possibility is to write books with the word "universes" in the title[8] and entertain the illusion that one has thereby escaped the burden of pondering whether there is a true totality of material things.

Consciousness should seem to embody precisely that foremost form of unity that is manifested in one's personal self-awareness. Exactly the opposite happens to a universe, which "is supposed to be everything," when it is endowed with consciousness through an abuse of quantum mechanics. The procedure flies in the face of that marvelous science on two counts. One is that quantum mechanics ceases to be itself if it is supposed to operate with a wave function which is not statistical. One still has to invent a statistics, quantum mechanical or other, that deals with unity. The other is the confusion of quantum mechanics with a most dubious philosophy. Whatever doubts one may have about idealism (Platonist or Hegelian) and pragmatism (operationism), their fusion can only result in the kind of philosophical monster which is gnosticism. It was precisely that monstrous perspective that was grafted on quantum mechanics by such major architects of it as Bohr and Heisenberg. The ultimate consequence of that gnosis is the claim that gluons, nuclei, atoms, and even the universe gain their reality when the wave functions representing them are made "to collapse" by mere thinking.

When thinking about mere abstractions, such as wave functions, is attributed the power to produce reality, including its totality or the universe, interest in rigorous proofs about the existence of that totality will yield to cavorting in *fiats*. Scientific cosmologists, who account for the universe by mere thinking about it, hardly think about questions such as these: Does the universe exist when the individual scientist does not think of it? Do individual scientists possess the ability to fuse their conscious moments into a single universal experience? Can they give assurance that in spite of the variety of their cosmological theories, the product of their thinking will be a scientifically structured universe valid not only for the centuries of science, but also for the two million or so years of human history that had preceded them? Have they provided the proof that a scientifically illiterate human consciousness, that had been around long before science appeared on the scene, can account for a universe demanded by science? Have they proved that long before the emergence of *homo sapiens* the "consciousness" of mammals, amphibians, fish, amoebas, and viruses had made the wave function of the universe "collapse" into a physical reality?

What sort of consciousness should one ascribe to the vast zoo of subatomic particles, if *they* had to achieve that cosmic feat long before the appearance of life on earth, or anywhere else in the cosmos? Does one escape the burden of proving the existence of the universe by resting the case with the consciousness of "extraterrestrials" who, as far as evidence goes, exist only on paper full of grossly overvalued probability formulas?[9] And even if they turned up, and spoke English to boot, would not the step, whereby one assigns to consciousness the task of producing the universe, be but a form of passing the buck on a cosmic scale? Or should one expect the final answer from claims about artificial intelligences that invariably presuppose what they were meant to demonstrate?

The task of proving the existence of the universe as an absolute totality of things loses its meaning in the multiworld theory. It still has to cope with the task of proving that there are *other* observers and thereby provide escape from a subjectivist straitjacket. Another and widely entertained form of the subjectivization of the cosmos is the strong form of the anthropic principle,[10] which again prompted, as one may expect, no efforts to prove the existence of the cosmos. To assume the existence of the cosmos, just because there are humans to observe it, should seem rank primitiveness when compared with the allegedly primitive reasonings of great philosophers from the geocentric age. From Aristotle to Thomas Aquinas, of whom more later, they may not have felt the need to prove the existence of a cosmos because they thought to see its boundaries in the sphere of the fixed stars.

Disinterest in the logical need to prove the existence of the cosmos, which is the totality of all material things, is further encouraged when the cosmologist abolishes the difference between being and non-being, or nothing. The first form of this 20th-century cosmic farce came, as was seen, with the steady-state theory. At any rate, it is not possible to give a proof of the existence of the universe within that theory for several reasons. For one, neither the steady-state theorists and their admirers, nor their original critics cared to note that the nothing cannot be invested with that observability which must, by definition, be the hallmark of scientific proof. The existence of nothing, which is a scientific part of that theory's definition of the cosmos, cannot be proven scientifically even if it implies the coming into existence of such a small bit of matter as the hydrogen atom. The forging of a proof of the existence of the universe is further undermined within that theory by that infinite amount of matter which it claims to generate. Such an amount does not posses countability, a property absolutely indispensable for the purposes of science.

It cannot deal with such contradiction in terms as embodied in an actually realized infinite quantity.[11]

Nor can a proof of the existence of the universe be formulated within the inflationary theory, this third principal modern cosmological theory wherein the universe disintegrates into countless universes. It became the dubious distinction of the proponents of that theory to bank in, on a cosmic scale, on a logical error or non-sequitur, taken for a revolutionary insight. The error was incubated, hatched, and spread by the Copenhagen interpretation of that splendid science which almost deserves Schrödinger's encomium of it as "the Lord's quantum mechanics."[12]

What is fundamentally bad in that interpretation is that it endorses a non-sequitur whereby the ontological turns into a mere function of the operational.[13] For it is a non-sequitur to claim that if an interaction cannot be measured exactly, it cannot take place exactly. This non-sequitur, or the jump from the operational to the ontological domain, is the gist of the philosophy which Heisenberg and Bohr grafted on to what is usually called the principle of indeterminacy, but should have rather been called the principle of imprecision.[14] Once that non-sequitur is accepted, no further objection can be made to the claim that the physicist can literally create matter, indeed an entire universe, out of nothing. And since the ontological distance between non-being and being is infinite, it is no greater a feat to create an entire universe than to produce a mere fraction of the atomic nucleus out of nothing.[15]

Such a fraction, perhaps of the order of 10^{-34} gm, is on hand when one tries to measure the moment or time of the emission of an alpha particle, say, from a radium atom. The measurement is subject to the uncertainty principle which can be written not only as $\Delta x \,.\, \Delta mv \geq \hbar$ (where x is position, mv is momentum and $\hbar$ is Planck's quantum divided by 2π), but also as $\Delta E \,.\, \Delta t \geq \hbar$ (where E is energy, t is time). But since E can also be written as mc^2, ΔE becomes Δm (or an error in measuring the mass of the alpha particle), because c or the speed of light is invariable. In itself all this means only that there is a limit to the precision of measuring Δm. However, something entirely different is meant if one grants the Copenhagen interpretation according to which what cannot be measured exactly, cannot take place exactly. Then Δm, or the error in measuring m, becomes a real or ontological mass defect.

Such a defect cannot be remedied by the methods of exact science, be it quantum mechanics. As a mathematical formalism, quantum mechanics is even more limited with respect to reality than, to paraphrase an old remark of T. H. Huxley, is a mill with respect to wheat when given, instead of grain, mere peascods to grind.[16] This elementary truth was recognized very recently by an advocate of the quantum creation of the universe, of which more later, as he mused: "Like most things in life, theoretical physics does not yield something for nothing, and what you get out is what you put in."[17] No science can deliver real matter, unless it is there, in the first place. Only in terms of an adulterated science (that is, a good science corrupted by bad philosophy), can one imagine that one can keep the books of physics balanced by stealing matter from "nothing." An unwitting espousal of bad philosophy is on hand when the scientific cosmologist accepts as authority a dictionary article by a professional philosopher and speaks of it as "an illuminating account of why the concept of 'nothing' is nothing to worry about."[18] Science is indeed in a sorry state when its cultivators take cover in writings of philosophers for failing to perceive an obvious fallacy in logic.

These are some of the points which some proponents of the inflationary theory of the universe fail to comprehend. This is especially true of A. H. Guth, the original proponent of that theory.[19] He has repeatedly emphasized that when he claims to create entire universes out of nothing, he meant a creation "literally out of nothing"[20] and not out of some recondite form of energy. Such an energy has to be "something" (the very opposite of nothing) if it is to be subject to the mass-energy equivalence and be part thereby of physics and not of some disreputable form of metaphysics. Guth's claim is undoubtedly the most extraordinary claim of the entire history of science. Yet when confronted with it, cosmologists failed to do even something which is purely scientific, though in this case the very essence of the scientific method is at stake. They did not protest against the introduction into science of what is strictly unverifiable empirically. The "nothing," to say nothing of what is "literally and absolutely nothing," cannot be the object of that empirical verification which is part and parcel of reputable science. Nor can a universe be thought of as truly existent if the "nothing" is an essential ingredient of it.

There is, of course, another reason why in terms of the inflationary theory one cannot give a consistent definition of the universe, let alone prove its existence. Being a statistical theory, it has to start with a statistically significant or very large number of embryo universes whose number cannot be specified in advance. Further, there is nothing in the theory that would assure a cut-off point in the process of deriving the set of these embryo universes from another statistical ensemble.

The situation becomes even more desperate from the logical viewpoint when statistical randomness is predicated about the set of physical laws governing the individual universes. Needless to say, no details have been forthcoming about another universe governed by a different law of gravitation, by a different kind of Coulomb force, by constants different from those in our universe, or about a universe without physical constants, or with constants that vary randomly and so forth. Nor has information been given so far about the ensemble from which all those universes spring forth. Is that ensemble an overall physical reality and therefore the Universe? Or is it merely a convenient red-herring, dressed up in mathematical garb, to distract anyone mindful of the impossibility of endlessly postponing a true starting point? No answer has been forthcoming to the question: What overall law turns those universes into a unit, that is, into a Universe? No wonder. The inflationary theory of the universe turns the Universe into a Humpty Dumpty, who, once he had fallen from the top of the wall, could not be put together again. In vain do the inflationary theorists try to save even conceptually that coherent totality which is the Universe.

At any rate, the unity which the Universe has to be cannot be the starting point of a statistical theory. For if it is true that "the interpretation of probability as applied to a *single* electron remains as mysterious as ever,"[21] the handling of the Universe in terms of the same probability functions should seem preposterous. What is at issue is not the formulation of a single initial quantum state for the universe, an enterprise to which much impetus was given in 1983 by J. B. Hartle and S. W. Hawking,[22] but that singleness which the Universe has to be if it is truly one. Even in respect to that "unique state-function" of the universe, efforts to specify it had posed "the danger that the original problem of the multiplicity of possible initial states will be replaced by an equivalent multiplicity to determine it uniquely!"[23]

As to those cosmologists who claim the constant flowing of new "worlds" into our universe defined by the horizon, they should at least reflect on the reason why those other "worlds" can fit into the one already observed. For if they "fit," it is only because they obey the same laws as the one into which they drift. This suggests that as far as science is concerned there can be no truly "other" worlds or universes. For if they are truly "other" universes, neither can they physically tie up with this universe, nor can they become the object of a meaningful (empirically verifiable) scientific theory.

This remark is an advance note about the object whose existence is to be proved in this essay. The object is not the observed universe, or the one within the horizon, not even the one which corresponds to any given space-time manifold. Thus apart from the all too cavalier handling by scientific cosmologists of the term "universe," it should be clear that the new science of cosmology is not necessarily a science of the actual physical universe as a strict totality of things.

No proof that such a totality does exist, can be found in books and articles on scientific cosmology. There at best one can find it stated that "the greatest supposition is perhaps that shared by all cosmologists, namely, that (i) the material universe exists and (ii) it is susceptible in its entirety to mathematical analysis."[24] Still, scientists should not be taken to task too harshly for not trying to prove the existence of that totality. Philosophers, whose business it would have been to formulate the proof, provided neither encouragement nor inspiration for scientists. These in turn failed to challenge philosophers, an old story by 1905, the year when Chesterton deplored the studied neglect of "everything."

If philosophers had taken note of what scientific cosmologists, including those busy with a "quantum creation" of the universe, had been discussing, Dirac might have had a different recollection about the philosophers' reaction to the science of relativity. Fascinated by newspaper reports about the verification of Einstein's prediction of the bending of starlight around the sun, Dirac, an engineering student then at the University of Bristol, enrolled in a course on relativity given by the philosopher, C. D. Broad. What he learned from that course, during which he encountered no mathematical formulas, was that "philosophers just put forward the view that everything had to be considered relative to something else, and they rather claimed that they had known about relativity all along."[25] Confronted with modern scientific cosmology, most modern philosophers would have found it difficult to say that they had always known of the universe. Its existence has, in fact, been for some time a chief target of their scepticism.

[1] In a conversation with Manfred Clynes, as reported by P. Michelson, *Einstein: Profile of the Man* (New York: Dodd, 1962), p. 253.

[2] H. Bondi, *Cosmology* (2d ed.: Cambridge: University Press, 1961), pp. 3 and 10.

[3] The concluding aphorism in *Tractatus Logico-Philosophicus*.

[4] Bondi, *Cosmology*, p. 143.

[5] For details, see my *God and the Cosmologists*, pp. 158-59. Chesterton's famous reply to an advocate of solipsism, "Cherish it!" is the appropriate remark which advocates of that theory deserve.

[6] P. Davies, *Other Worlds: A Portrait of Nature in Rebellion. Space, Superspace and the Quantum Universe* (New York: Simon and Schuster, 1980), pp. 134-35.

[7] M. Kafatos and R. Nadeau, *The Conscious Universe: Part and Whole in Modern Physical Theory* (London: Springer Verlag, 1990). Contrary to the impression given by the main title of the book, the universe as such is nowhere discussed in it and its combined name and subject index contains no entry under 'universe'. The book, which therefore also contradicts its subtitle, offers the customary potpourri of Eastern mysticism, quantum mechanics, Kantianism, and, above all, an invariably vague use of basic philosophical terms.

[8] See, for instance, J. Leslie, *Universes* (London: Routledge, 1989), where poor philosophizing is burdened with scientific amateurism.

[9] In evaluating Drake's formula for the probability of intelligent life with advanced civilization on other planets, little attention is paid to the enormous improbability of at least two factors. One is the occurrence of tidal basins on earth, which are largely the result of lunar tides, and these in turn should appear most unlikely in view of the extreme improbability of a duplication of the earth-moon system. The latter is explainable only in terms of a collision between the proto-earth and an intruder into the planetary system, both with a large number of specific parameters. For details, see ch. 7, "Earth's Luck," in my *God and the Cosmologists* (Edinburgh: Scottish Academic Press; Washington, D.C.: Regnery Gateway, 1989). The other is the development of science on earth which should seem a most unexpected result in view of the invariable stillbirths of science in all great ancient cultures. Even more improbable should seem the rise of a viable science in view of a most unpredictable factor needed in that respect which is the Incarnation as a safeguard of the idea of creation out of nothing and in time, on the basis of which the idea of inertial motion was formulated more than three hundred years before Newton. For details, see ch. 2, "The Birth that Saved Science," in my *The Savior of Science* (Washington D. C.: Regnery Gateway, 1988).

[10] It asserts much more than the weak form of the principle, according to which the universe had to have, from the start, very specific characteristics if a carbon-based human organism was to emerge eventually.

[11] This difficulty of the theory, emphasized in my *Cosmos and Creator* (Edinburgh: Scottish Academic Press, 1980), is systematically ignored in references to it.

[12] E. Schrödinger, *What Is Life? And Other Scientific Essays* (Garden City, N.Y.: Doubleday, Anchor Books 1956), p. 83.

[13] This is the formulation in which I have kept calling attention, now for over a decade, to the basic fallacy of the Copenhagen interpretation. See my article, "Chance or Reality: Interaction in Nature versus Measurements in Physics" (1981), which is the first chapter in my *Chance or Reality and Other Essays* (Lanham Md.: University Press of America, 1986) and my essay, "Determinism and Reality," quoted below in note 14. The failure of my efforts to make a noticeable dent on the "consensus" duplicates the manner in which the warning of E. Turner, professor of philosophy at the University of Liverpool, made in the pages of no less an important scientific weekly than *Nature*, fell upon deaf ears among physicists as early as 1930. In his letter to the Editor, Turner wrote: "Every argument that, since some change cannot be 'determined' in the sense of 'ascertained,' it is therefore not 'determined' in the absolutely different sense of 'caused,' is a fallacy of equivocation" (Dec. 27, 1930, p. 995).

[14] As emphasized in my *God and the Cosmologists*, pp. 118-19. See also my essay, "Determinism and Reality," in *Great Ideas Today 1990* (Chicago: Encyclopedia Britannica, 1990), pp. 277-302.

[15] That this point can be wholly lost on the modern physicist or cosmologist is best illustrated by a claim of D. Sciama (quoted in *Origins*, p. 142; see note 27 to Chapter 2) that "continuous creation of matter is even less a of thing to introduce than the creation of a whole universe at one go."

[16] T. H. Huxley, *Lay Sermons: Addresses and Reviews* (5th ed.; London, 1874), p. 249.

[17] C. J. Isham, "Creation of the Universe as a Quantum Process," in R. J. Russell, W. R. Stoeger, G. V. Coyne (eds.), *Physics, Philosophy and Theology: A Common Quest for Understanding* (Vatican Observatory: Vatican City State, 1988), p. 402.

[18] Ibid., p. 406, note 7. The article referred to is by P. Edwards in the *Encyclopedia of Philosophy* (New York: Macmillan, 1967), vol 6, pp. 524-25.

[19] Guth's own, non-technical, version of that theory is Chapter 5, "Starting the Universe: The Big Bang and Cosmic Inflation," in *Bubbles, Voids, and Bumps in Time: The New Cosmology*, ed. J. Cornell (Cambridge: University Press, 1989).

[20] See Guth's article, co-authored with P. J. Steinhardt, "The Inflationary Universe," *Scientific American* April 1984, pp. 116-28, and his interview in *Origins*, where he contrasted his views about the origin of the universe before and after his formulation of the theory: "I was *not* thinking about

how the universe was created. . . . At that time I was not at all aware that it was possible to make a universe from nothing. I would have thought at that time that conservation of energy would make that impossible. Now, I think it's very plausible that the universe is a quantum fluctuation, starting from *absolutely nothing*" (p. 476, italics added). Guth failed to realize that those who had formulated the principle in question saw its connection with the more basic principle of the conservation of matter, which, from the very start of its formulation, rested on the realization of an ontological principle, a principle for which Guth shows no respect at all.

[21] C. J. Isham, "Creation of the Universe as a Quantum Process," p. 404.

[22] J. B. Hartle and S. W. Hawking, "Wave function of the Universe," *Physical Review D* 28 (1983), pp. 2960-2975.

[23] A truly illuminating remark of Isham in his "Creation of the Universe as a Quantum Process," p. 405.

[24] Ibid., p. 401.

[25] See H. Kragh, *Dirac. A Scientific Biography* (New York: Cambridge University Press, 1990), pp. 5-6.

4

Orphaned by Philosophy

What has been said so far should make appear forlorn the hope that philosophers would be challenged by scientists to think hard about the universe. But even if such a challenge had been forthcoming, it would have hardly found significant echo among contemporary philosophers. They are heirs to two main trends—empiricism and rationalism—within both of which the universe was handled as an orphaned child unable to attract proper attention.

The empirical trend has no better starting point than William Ockham. This may seem paradoxical because Ockham's own starting point was theological. But since the starting point was extreme, it invited the other extreme, amply illustrated by subsequent empiricists, where both theology and cosmology ceased to have intellectual appeal. As one of the "spiritualist" branch of Franciscans, Ockham nurtured a quasi-mystical urge to see God's direct action everywhere, an urge that made him see each event as a miracle performed by God himself. Ockham sought a philosophical justification for this miracle mongering by denying to things the ability to function as secondary causes. He thereby abolished natures as consistent entities with appropriate effects. He indeed explicitly claimed that it was possible to have starlight without stars.[1] The abolition of individual natures brought him dangerously close to abolishing Nature, or the Universe. He certainly rejected the argument widely entertained by his medieval predecessors that the unity of the universe was due to an internal connection among all things. According to Ockham, the unity of the universe was merely the circumstance that "some things stand farther apart from one another than others."[2]

Such was one of Ockham's very few utterances on the universe. Their nominalist logic is reflected in Francis Bacon's coolness toward Copernicanism,[3] in Hobbes' crude distinction between the universe and the impossibility to determine anything about it,[4] in Locke's diffidence about anything beyond the solar system,[5] and in Hume's derision of the universe as typical of products concocted in spiders' entrails.[6] The same is true of Auguste Comte's characterization of astrophysics as a sort of insanity[7] and Mach's effort to erect an unbridgeable gap between the parts and the whole of the universe. Mach's claim that "the universe is like a machine in which the motion of certain parts is determined by that of others, only nothing is determined about the motion of the whole machine"[8] received its rebuttal in the expansion of the universe. Similarly devastating rebuttal was served by the cosmology of general relativity to Mach's subsequent contention that no part of the observable world can be taken for a representative sample of the universe.[9]

The revival of empiricism in the early 20th century did not promote interest in the universe as such. John Dewey was consistent with his basic and early premises as he declared late in his career that "the world as we experience it is a real world. But it is not in its primary phases a world that is known, a world that is understood, and is intellectually coherent and secure. . . . Knowledge then does not encompass the world as a whole."[10] About the same time, in the early 1930s, when observations of the recessional velocity of galaxies began to unveil cosmic spaces of billions of light years, P. W. Bridgman, the most thematically empiricist among major modern physicists, decried as useless all astronomical research beyond a few thousand light years.[11]

But just as those cosmic depths seemed illusory to Bridgman, so did the universe. He took the Copenhagen interpretation of quantum mechanics for a proof that "the world is not a world of reason, understandable by the intellect of man, . . . the world is not intrinsically reasonable or understandable." If such was the case, it was logical to call in doubt the existence of the universe as well. Bridgman implicitly did this as he declared in the same breath: "A vision has come to the physicist in this experience which he will never forget; the possibility that the world may fade away, elude him, and become meaningless because of the nature of knowledge itself." As if to insult the intellect after having injured it, Bridgman took this fearsome possibility for a source of genuine intellectual humility: That possibility, he added, "has never been envisaged before, at least by the physicist, and this possibility must keep him forever humble."[12]

That such a humility may hide the pride of scorn could be gathered from a remark of Bertrand Russell, made about the same time, in 1931 to be exact, at the high point of his empiricist phase. He debunked the universe as an entity which is "all spots and jumps, without unity, without continuity, without coherence or orderliness or any of the other properties governesses love."[13] The beginning of Bertrand Russell's empiricist phase was marked with a similar broadside against the universe. In his famed essay, "A Free Man's Worship," he celebrated the inevitable burial of "the whole temple of Man's achievement . . . beneath the débris of a universe in ruins."[14] Ironically, this essay saw wide circulation from 1917 on when Einstein crowned his achievement in general relativity with an essay on its cosmological consequences. As if to make the irony complete, Bertrand Russell declared in the same year, and in an essay entitled, "On Scientific Method in Philosophy," his firm belief that the conception of the universe was "a mere relic of pre-Copernican astronomy."[15]

Bertrand Russell did not care to retract any of these remarks as he admitted, in his famed radio-debates with Father Copleston in 1950, that the universe was a brute fact which "was just there, and that's all."[16] He did not try to explain why one could be so sure about this greatest of all facts. Perhaps he felt that by 1950 the weight of scientific cosmology was too much to take the universe too lightly. Nor did he renounce the pragmatist philosophy of William James in which he had found, around 1917, the basis for pouring scorn on the unity of the universe as well as on its reality. Almost at the start of the essay, "On Scientific Method in Philosophy," there stood a long quotation from William James' *Some Problems of Philosophy*, where the oneness of the universe is dismissed in phrases loaded with inner contradictions, none of which was noticed by the great mathematical logician that Bertrand Russell was.

For it was one thing to speak, as William James did, of things and worlds so different as "the dreams of different persons" can be and indeed worlds "so unlike and incommensurable, and so inert toward one another, as never to jostle or interfere." One could imagine that "even now there actually be whole universes so disparate from ours that we who know ours have have no means of perceiving that they exist." But it was another to state that whatever their being "so unlike and incommensurable," it was still possible "to conceive their diversity; and by that fact the whole lot of them form what is known in logic as 'a universe of discourse'."[17]

Neither Bertrand Russell nor William James suspected that by emphasizing both differences and a common discourse about them they were laying a minefield in logic. For unless those differences contained something in common, it was not logically possible to have a discourse that connected them together. If they did, they formed a universe and not merely a "universe of discourse" lacking logical coherence. Not to see this point was equally reprehensible on the parts of both, though for very different reasons. The fault of Bertrand Russell lay in his being a logician, whereas William James defaulted through offering the most sustained discourse about a pluralistic universe, this most seductive of all empiricist efforts to devour the universe and still have it.

William James changed from a physician into a champion of scientific psychology before he gave almost exclusive attention to philosophy, at the center of which stood his resolve to oppose monism in any form. In doing so he merely exchanged logical forms of monism with an illogical one which he called pluralism. In the process he dismembered the universe and did so a mere four years after Chesterton had deplored the universal indifference to that "everything" which is the universe. William James performed this dissection of the cosmos as he delivered the Hibbert Lectures at Oxford during the academic year of 1908-09 and published them, in short order, under the title, *A Pluralistic Universe.*[18]

The Hibbert Lectures were a dubious performance, though consistent with the Lowell Lectures which William James had delivered during the academic year, 1906-07, at Harvard University. Only a thorough misunderstanding on his part of what Chesterton stood for could make him begin with an accolade of Chesterton's "admirable collection of essays," *Heretics*. Had Chesterton ever taken note of the Lowell Lectures, published under the title, *Pragmatism*, he would have found more than enough grist for his sharply grinding mill in the book's subtitle, *A New Name for Some Old Ways of Thinking*.[19] Indeed, it was but a new name for an old misconception invented for the purpose of finding truth in action with a systematic disregard of an even older concern for saving truth from the pitfalls of scepticism.

What Chesterton, perhaps the foremost champion of the Universe in this century,[20] would have especially taken for an offense was that William James offered the phrase, "I think with Mr Chesterton in this matter,"[21] after he had begun his *Pragmatism* with the following quotation from *Heretics*: "There are some people—and I am one of them—who think that the most practical and important thing about a man is still his view of the universe. We think that for a landlady considering a lodger it is important to know his income, but still more important to know his philosophy. We think that for a general about to fight an enemy it is important to know the enemy's numbers, but still more important to know the enemy's philosophy. We think the question is not whether the theory of the cosmos affects matters, but whether, in the long run, anything else affects them."[22]

Chesterton would have been fully justified in taking a spirited exception to William James, who held high the idea of a "pluralistic universe" and was careful to distinguish it from the monists' "block universe."[23] Yet William James did not care to consider that his disagreement with the monists was as much a half-truth as the "universe" of the monists was a farce about the whole loaf. But the error of the monists in reducing the realm of the spiritual to the physical universe left intact the truth that the physical universe had to be fully consistent, that is, coherent, in order to be useful for scientists. William James' effort, however praiseworthy, to save the realm of the psyche from the teeth of physicalism, undermined itself by making the universe useless for physicists. The latter run the risk of becoming mere tinkerers in a corner of the cosmic whole once they take William James' pluralism at face value.

Whatever the mechanistic or monistic philosophies grafted on Newtonian physics, the validity of Newtonian physics, and of all useful physics for that matter, depended on what William James stated about the monists. In his words, they insisted that "when you come down to reality as such, to the reality of realities, everything is present to *everything* in one vast instantaneous co-implicated completeness—nothing can in *any* sense, functional or substantial, be really absent from anything else, all things interpenetrate and telescope together in the great total conflux."[24] In a genuinely scientific universe, governed by universally valid laws, everything had to be seen as interacting, however remotely, with everything all the time.

William James failed to consider the precious little which scientists, let alone their cosmologist kind to come, could do with the pluralistic universe whose gist he gave in a graphic image. He spoke of a log "carried first by William and Henry, then by William, Henry, and John, then by Henry and John, then by John and Peter and so on." What he meant to say was that "all real units of experience overlap."[25] Later, without propping up the supposed discontinuous continuity of things acting upon other things with a reference to persons dealing with things, William James set forth the same idea: "Without losing its identity a thing can either take up or drop another thing like the log I spoke of, which by taking up new carriers and dropping old ones can travel anywhere with a light escort." Compared with this "pluralistic" universe, where connections and disconnections could readily alternate in any and all processes, in the monistic universe "everything, whether we realize it or not, drags the whole universe along with itself and drops nothing. The log starts and arrives with all it carries supporting it. If a thing were disconnected, it could never be connected again."[26]

Whatever confirmation William James thought to derive from the science of psychology on behalf of his elastic cosmic view, he could find little support in the hard sciences of his time. Of course, already around the turn of the century some scientists, who waxed philosophical in their non-scientific hours, began to talk of chance as if it were more than a cover-up for ignorance. Yet even long after the advent of quantum mechanics little light has been generated about the scientific use of chance and even of probability. This is particularly true of sanguine references to "chaos," this latest pseudo-sophistication in terms of which randomness is being nowadays celebrated by scientists and science-writers.[27] They may have been more circumspect if alerted to the fact that the question, "What *is* chance?," demands an ontological answer. Yet their reaction may have been the same baffled expression which a prominent champion, now dead, of the Copenhagen interpretation gave to this writer who pressed him for such an answer to that question.

Most scientists, as will be seen in the next chapter, have not even taken a close look at their inability to think clearly about chance or random events even as targets of a purely mathematical formalism. They would today blithely assure William James that the Copenhagen interpretation of quantum mechanics has put the supreme scientific seal on his idea of a "pluralistic" universe. They would indeed find, as will be seen shortly, among William James' utterances some that come all too close to a major consequence of the Copenhagen interpretation, or to taking the existence of external things for a function of conscious thinking about them. In whatever sense William James might today challenge some scientists to do clear thinking, many post-Einsteinian cosmologists would, with their commitment to the Copenhagen interpretation of quantum mechanics, find William James' idea of a "pluralistic" universe a proposition to be espoused rather than to be challenged.

At any rate, even apart from the crucial issue of the true nature of relation between things, if these were utterly conditioned by our thoughts about them, an issue not confronted by William James, he did his best to put together what he had first fragmented. He insisted that "our 'multiverse' still makes a 'universe'; for every part, tho it may not be in actual or immediate connection, is nevertheless in some possible or mediated connection, with every other part, however remote, through the fact that each part hangs together with its very next neighbors in inextricable interfusion."[28] He himself realized that in saying this he came all too close to endorsing the idea of a "block universe," an inference against which he spiritedly defended himself.

His last words on the subject show the existential need for a philosopher to retain the unity of the universe whatever else he might wish to take it for: "This world *may*, in the last resort, be a block-universe; but on the other hand it *may* be a universe only strung-along, not rounded in and closed. Reality may exist distributively just as it sensibly seems to, after all. On that possibility I do insist."[29] He also insisted that the pluralistic universe is essentially "unfinished," that "total oneness [was] never getting absolutely complete," whereas the "block-universe" was "timeless and eternally complete."[30]

Whether this disjunction represented the truth about the universe may be left aside. From the purely logical viewpoint it should be seen as leaving much to be desired. And the same holds true of what William James stated in the very same context about philosophy: "Philosophies are intimate parts of the universe, they express something of its own thought of itself. A philosophy may indeed be a most momentous reaction of the universe upon itself."[31] Did not William James come thereby very close to attributing the kind of consciousness to the universe which could not be a unitary consciousness but only a pluralistic one?

William James' suggestion of an absence of real difference between thoughts and things, individuals and universe, is a dubious heritage of modern philosophy which, in one way or another, tried to be "scientific." Particularly emphatic was this effort in the other main trend of modern philosophy, rationalism. All its chief proponents claimed to know that there was a universe, but the claim was in each case self-defeating, or totally useless from the scientific viewpoint, whatever that was at a particular time.

About Descartes' cosmology no less a 17th-century physicist than Huygens declared that it was a mere romance.[32] Huygens could have also found an interesting target in Descartes' hesitation about declaring the universe to be infinite. If extension was the basic parameter in reality, what could set a limit to it and thereby to the universe? Moreover, could not a Cartesian claim that he knew what the universe was with as great a certainty as the Creator did? Descartes, as is well known, was most hesitant to ascribe freedom to the Creator and came very close to stating that the world necessarily had to be created in the way in which he specified this. Clearly, if anyone a Cartesian rationalist could find no problem in answering the question: Is there a universe?

That such was indeed the inner logic at work within Cartesians became all too clear in Spinoza, a mind unencumbered either by Jewish or by Christian sensitivities about the prerogatives of the Creator. A central thesis of Spinoza's system is that "from the necessity of divine nature must follow an infinite number of things in infinite ways—that is, all things which can fall within the sphere of infinite intellect."[33] Lest anyone should misunderstand what he meant, Spinoza added: "From God's supreme power or infinite nature, an infinite number of things—that is, all things have necessarily flowed forth in an infinite number of ways, or always follow from the same necessity; in the same way as from the nature of a triangle, it follows from eternity and for eternity, that its three interior angles are equal to two right angles."[34]

In all appearance the universe was not a topic when, between November 18 and 21, 1675, Spinoza had several long conversations with Leibniz, who returned from Paris to Hanover via Amsterdam.[35] Had that topic come up, perhaps Spinoza could have given an advance hint of Leibniz's preference for an infinite universe, an idea which he was to put in writing forty years later, in his debate with Samuel Clarke on Newton's philosophy. There Leibniz claimed that the real philosophical task was to reconcile the finiteness of the universe with God's omnipotence rather than the other way around.[36] In the *Monadology*, which he wrote about the same time (1714), he stated that although "there is an infinity of possible universes in the idea of God . . . only one of them can exist."[37] He did not care to work out a proof that a universe indeed existed, although the question of its existence lurked between each and every line he devoted to his favorite brainchild, the pre-established harmony of all.

Leibniz's speculations about the monads, each of which mirrored, to recall his expression, the whole or the universe,[38] should again have led him to a thematic discussion of the question whether there was a whole. He exemplified the rationalist who is so much taken up with a particular aspect of reasoning as to forget about reason itself. Rationalism as it flourished during the 18th century showed many examples of this and also of the inner dynamics whereby such one-sidedness invites its very opposite, or materialism.

That modern materialism is an offshoot of Cartesian rationalism is demonstrated by the fact that 18th- and 19th-century rationalists gave much the same answer to the question, "Is there a universe?," which 19th-century materialists were to give. They all claimed to know what the universe had to be: infinity incarnate. While the French encyclopedists still had misgivings, perhaps for fear of the royal censor, about endowing the universe with infinity,[39] 19th-century rationalists had no hesitation. What Hegel said on the right, the materialists, including their dialectical brand, stated on the left. "The world surrounding us is an infinite whole, composed of the same materials, and moved by the same forces," declared L. Büchner, a medical doctor, in his runaway bestseller, *Kraft und Stoff*. His scientistic dogmatism reached its natural peak as he also claimed that "the immortality of matter is now a fact scientifically established, and can no longer be denied."[40] According to E. Haeckel, the next major German high-priest of scientific monism, "the extent of the universe is infinite and unbounded; it is empty in no part, but everywhere filled with substance."[41]

It was with the same assurance that Herbert Spencer discoursed about the evolution of the universe from a primordial homogeneous nebula to its actual specificity, without realizing the illogicality of this procedure.[42] For none of these brave minds did the question, "Is there a universe?," pose any problem. They answered the question about the existence of the universe by giving their reasons why it had to be what they thought it was and could not be anything else.

One, and not a small, problem with these brave minds was that they were wrong scientifically on more than one count. From the mid-19th century on it took a sort of defiance if not plain blindness to ignore the gravitational and optical paradoxes of an infinite universe. As to cosmogenesis, it was also in trouble in the measure in which Laplace's theory of the evolution of the solar system revealed its fatal shortcomings, the chief of them being its inability to cope with the distribution of angular momentum in the solar system.[43] Twentieth-century rationalists, idealists, and materialists, or any mixture of these very dogmatic philosophies, preferred to stick to generalities whenever the universe as such became the topic. Classic illustrations of this are the systems proposed by Bergson, Alexander, Whitehead, and Popper. They all answer the question, "Is there a universe?," by claiming to know, in terms of artfully generic statements, the very process whereby the universe takes on, during its eternal duration, all possible forms of existence. What they do is evade, by using an unspecified notion of "all," the question about that specific "all" which is the universe.

Particularly baffling should seem that claim, which cannot be verified at all, when it comes from a major philosopher who first excelled in science where the principal test of truth is empirical verification or observation. Yet the claim is central in what was called the last major philosophical cosmology, A. N. Whitehead's *Process and Reality*, first published in 1929. Judging by its subtitle, "An Essay in Cosmology," one at least of its twenty-eight chapters, if not one of its five Parts, should have dealt with the cosmos or universe. Far from it. Even the last chapter, "God and the World," is far more about a pantheistic God, who co-acts and even co-suffers with the world, than about the world itself. None of the half a dozen references in that book to relativity are to its cosmology. With his fleeting reference to "the immensity of the world," Whitehead merely wanted to negate "the belief that any state of order can be so established that beyond it there can be no progress."[44]

In saying this, Whitehead gave away something of his debt to Bergson and revealed much of the self-defeating tactic whereby purpose and humanness are reclaimed through a vitalization of the cosmos. Bergson found the root of all scepticism about purpose in considering the universe as a strict totality. He therefore had no choice but to fault eternalism as well as creation in time: "Whether we speak of creation or posit an uncreated matter, it is the totality of the universe that we are considering at once. . . . As thinking beings we may apply the laws of our physics to our own world, and extend them to each of the worlds taken separately; but nothing tells us that they apply to the entire universe, nor even that such an affirmation has any meaning; for the universe is not made, but is being made continually. It is growing, perhaps indefinitely, by the the addition of new worlds."[45] Such was Bergson's artfully ambiguous but in essence very unhesitating answer to the question whether there was a universe. Such an answer is neither scientific nor philosophical but quasi-mystical. It was no accident that Bergson defined the universe as a "machine for the making of gods."[46]

The author of *The Open Universe and Its Enemies* gave a scientific coating to the Bergsonian phrase, "open universe,"[47] and followed both Bergson and Whitehead in slighting modern scientific cosmology. Karl Popper's first major work, *The Logic of Scientific Discovery,* contained nothing on the greatest of all scientific discoveries, or the first scientifically consistent treatment of the totality of gravitationally interacting things in a central field of force subject to the inverse square law. He "improved" on this strange performance in his last major work, *The Open Universe: An Argument for Indeterminism*, by directly slighting scientific cosmology. All too eager to refer to the latest doubt on the distance-velocity relation, Popper spoke about the "precariousness of the scientific status" of the cosmology of the Big-Bang.[48] In view of this Popper may have offered an unintended slur on science in making his often quoted statement that "all science, I believe, is cosmology."[49]

As the proponent of the theory of "three worlds," Popper was not to advocate their unity in a well-defined World or Universe. His advocacy of indeterminism had for its target not only the closed, or mechanistic, world of Laplace but also any closed world, however probabilistic: "We live in an open universe. We could not make this discovery before there was human knowledge. But once we have made it there is no reason to think that the openness depends exclusively upon the existence of human knowledge." If, however, this was the case, the universe too was alive. Bergson would have noted that Popper differed from him merely in semantics as he spoke of emergence instead of *élan vital*: "Our universe is partly causal, partly probabilistic, and partly open: it is emergent."[50] Once more, as so often in the past, the turning of the universe into a quasi-living entity undercut the possibility of facing up to the task of proving the existence of a strict, all-encompassing totality of things.

The universe certainly did not fare well in the hands of logical positivists who successfully claimed to themselves, for over three or four decades, the aura of scientific rationality. This should seem a foregone conclusion to anyone mindful of the veneration which the thought of Ernst Mach enjoyed in the Vienna Circle. Mach still tried to escape the consequences of his reduction of the world to sensations of colors, tones, and pressures and save thereby its objective existence. As a physicist, who daringly ascribed the inertia of bodies to the interaction of all matter in the universe, he obviously needed a matter that was more than one's sensations and far more than the sensations of all men taken together. But as one obsessed with the elimination of all difference between physics and psychology, hardly an obsession generated by the practice of physics, he had to say that "matter for us is not what is primarily given."[51] No wonder that Mach the cosmologist could never speak of the universe as something given.

Nor could the logical positivists, who did not even care for the universe which Einstein's science was unfolding under their very eyes. That there was nothing about the scientifically articulated universe in R. Carnap's "logical construction of the world," foreshadowed, since its publication in 1928,[52] the attitude of logical positivists to questions of cosmology. A year later, in H. Reichenbach's *Atom und Kosmos* all discourse about the cosmos or universe was restricted to that word on the title page.[53] Thirty years later Reichenbach improved on his silence about the universe only to the extent of saying that "we have no absolutely convincing evidence that there is a physical world." The only saving grace in this dictum was the consistency with which he continued: "We have no absolutely convincing evidence that we exist."[54]

None of the logical positivists was as consistent as O. Neurath. He found the very title, "Vienna Circle for the Dissemination of the Scientific World Outlook," of the Circle's Manifesto potentially misleading because of the presence of the word "world" there: "It would be less misleading to speak of the 'Vienna Circle for Physicalism'," he wrote in 1931, "since 'world' is a term which does not occur in the language of science."[55] Yet all logical positivists were consistent to the point of eschewing the science of the universe made possible by Einstein whenever they turned their discourse to his contributions. The best illustration of this is A. J. Ayer's last major book, *The Central Questions of Philosophy*, where the detailed index contains no references to "universe" and "world."[56] No wonder. Ayer's ultimate aim was to discredit once more natural theology, an enterprise which should seem very easy if that entity, which is the World or Universe, is studiedly ignored. The enterprise readily turned into a boomerang. Had Ayer paid serious attention to the universe, he would not have had to admit, a few years later, on the BBC that the "most important of the defects [of logical positivism] was that nearly all of it was false."[57]

Such is an all too obvious registering of a total debacle or shipwreck that allows its victims to reclaim but insignificant pieces from the disaster. When this happens in connection with a philosophical construct one may look for a pattern evident in the collapse of huge edifices. Almost invariably the cause of the trouble is some basic fault of the architect's design. In philosophy that basic fault relates to the initial step in reasoning, a step that is never made without an ultimate aim in mind. The fondness of logical positivists for a purely "rational" outlook on existence was a piece of their profound dislike for anything transcendental and especially for that transcendence which is God. Thinking about God fared indeed very poorly in Ayer's Gifford Lectures, where a full half a century after Einstein created a new branch of science, cosmology, the universe was hardly visible at all.[58]

In all the rationalist trend of modern philosophy, Kant is the only one who saw the central role of the universe in any discourse with some claim to being metaphysical, however rationalist. The widely shared conviction that Kant took the genuinely metaphysical sting out of the universe, may be the reason why references are few and far between to the centrality of the universe in Kant's rationalist philosophy. Only on occasion can one find in the philosophical literature contemporaneous with the explosive growth of modern scientific cosmology a proper exposure of Kant's essential aim: "Kant expressly argues that since it is unintelligible to suppose that the world is *either* finite *or* infinite, it must be the case that no objective world exists! Kant is the first philosopher in whom the world comes to sight as a problem simply as world."[59] Neo-Kantian interpreters of Einstein's relativity have as a rule ignored the problem posed by his cosmology to Kant's pivotal claim about the intellectual status of the idea of the universe.[60]

Rather unusual is therefore the sustained effort of M. Munitz to refocus on the merits of Kant's contention about the universe, an effort which began with his *The Mystery of Existence*.[61] There Munitz reargued what Kant had already argued, namely, that the answer why the universe does exist cannot be given because the universe, or the totality of things, cannot be known unambiguously. Munitz set himself the task of ascertaining whether Kant had left something really important out, but he could find nothing of the sort. Nor did he find anything new in the formulation of the cosmological argument after surveying modern philosophy and modern scientific cosmology, and last but not least, such modern Christian formulators of the cosmological argument as Copleston and Mascall. Curiously, Munitz did not note that neither Copleston nor Mascall gave as much as a short proof about the existence of a strict totality of things.

Munitz distinguished between two basic meanings of the word "universe." One was "that to which the observed universe belongs," or, if I interpret him properly, an entity defined by the actual horizon of observability, a horizon that can extend. The other was "that which is implicitly defined by the traits assigned to it by some cosmological model." Then he added that the idea of absolute totality might be conceivably grafted on the second definition but obviously not on the first. His reason was that nothing guarantees that the wider universe which is beyond the horizon is "pervasively structured," that is, ordered.[62] He further argued that while the second meaning of the universe may include the meaning of an absolute whole, it remains a mere model which as such cannot be verified empirically. According to Munitz, the realist (by whom he seemed to mean a Platonist), is ready to identify the model universe with the observable, but he does so on the basis "of a metaphysical faith" and not on an empirically testable ground.[63]

Munitz then turned to the idea of "absolute totality" which he found wanting for two reasons. One is that "absolute totality" is not sufficiently clear because its conceptual contours cannot be delineated against something else. Second, he found no warranty for assuming that the absolute totality is "pervasively ordered," that is, has the same structure throughout and a structure identical with the already explored part of the universe.[64] (The stupefying consequences of this for scientific cosmology should seem obvious, though Munitz tried to sidestep them even a quarter of a century later[65]). Munitz therefore held the idea of universe as "absolute totality" to be an age-old tendency of projecting "what is a human need into an allegedly antecedent ontologic fact."[66]

It should be observed, parenthetically, that Munitz is wrong in thinking that the idea of cosmos "as an ordered totality that binds all phenomena in an universal scheme, and whose details are accessible to patient rational inquiry" is a bequest of the Greeks of old to human civilization.[67] For all

Greek philosophers, and notably for the greatest—Plato and Aristotle—among them, the universe was partly ordered, partly disordered. Emphatic insistence on the full orderliness of the universe first appears only centuries later, in the anti-Arian writings of Athanasius. It was he who claimed that a fully ordered universe could alone issue from the creative power of a fully divine, and therefore infinitely rational, Logos.[68]

At any rate, it should not seem surprising that Munitz upholds, with Hume, the distinct possibility of disorderly parts in the realm of existents. And just as Hume, Munitz too fails to define what he means by disorder. He does not seem to be bothered by the contradictory character of total disorder or absolute chaos. His view, that "Kant showed convincingly that all these notions of a unified system of thought express, at best, a *regulative idea* of inquiry,"[69] seems to suggest that therefore this idea should be taken seriously. One could only wish that Munitz had emulated the consistency of Schopenhauer who wrote about Kant's reclaiming the universe at least as a regulative idea: "Kant can barely have been serious in making that assertion."[70]

In view of what is widely known about Hume's "spiritual" odyssey, it is difficult to avoid the conclusion that he had deliberately constructed a reasoning which leads to the impasse of agnosticism or rather scepticism, with discourse on the universe becoming a chief victim. Much less is remembered the catalytic effect which Rousseau's *Julie* had on Kant's latent atheism and on his subsequent resolve to give a rational foundation to Rousseau's romantic emancipation of man from all transcendental constraints. For a glimpse of Munitz's motivations, we must turn to his *Cosmic Understanding,* where he portrays man standing at the intersection of Boundless Existence and the Universe, unsure of either.

Concerning one's uncertainty about the universe, Munitz offered no new arguments. He, however, brought fresh perspectives with his frank declaration that Boundless Existence, by which Bergson, Whitehead, Popper and others had set so great a store, "is a surd. It ultimately defeats . . . all efforts at rationality." No less significant is the candidness with which he drew the supreme moral: "We should therefore take life with less than total seriousness. We can 'smile' at it and take it 'lightly.' And this combination of seriousness and lightness is what underlies the serenity born of an intensified awareness of Boundless Existence. In the degree in which we can succeed in achieving this balance of attitudes is born one important avenue to the peace of spirit possible to man."[71] Whether that avenue is open to the overwhelming majority of mankind caught in miseries of all sorts should seem most doubtful. What is above any doubt is that along that avenue one cannot reach that foremost object which is the Universe.

It should therefore be clear that a starting point other than the ones offered by rationalism and empiricism must be chosen if one is to find real instructiveness in talking about the universe as a strict totality of things. Such an instructiveness must in turn include the proof that there is a Universe.

[1] For such and similar details, see E. Gilson, *The Unity of Philosophical Experience* (New York: Charles Scribner's Sons, 1937), pp. 67-91. Today Ockham would claim that it is possible to have an electric field without electric charges.

[2] See the word "mundus" in L. Baudry, *Lexique philosophique de Guillaume Ockham* (Paris: Lethielleux, 1958), pp. 163-64.

[3] Particulary revealing in this respect is Bacon's "Descriptio globi intellectualis."

[4] See especially the chapter, "Of the World and of the Stars," in Hobbes' *On the Body* where he wrote: "Concerning the world, as it is one aggregate of many parts, the things that fall under inquiry are but few; and of those we can determine none." Then he added, in blissful oversight of contradicting himself: "Of the whole world we may inquire what is its magnitude, what is its duration, and how many there be, but nothing else." See *Hobbes. Selection*, ed. F. J. E. Woodbridge (New York: Charles Scribner's Sons, 1930), p. 125.

[5] In order to become "certain that there is a God . . . we need go no further than *ourselves*," wrote Locke in *Essay concerning Human Understanding* (New York: Dover, 1959), vol. 2, p. 307.

[6] Especially in his *Dialogues concerning Natural Religion.*

[7] A. Comte, *Cours de philosophie positive* (Paris: Bachelier, 1841), vol. 2, p. 2.

[8] E. Mach, *History and Root of the Principle of the Conservation of Energy* (1872), tr. P. E. B. Jourdain (Chicago: Open Court, 1911), p. 62.

[9] See his *The Science of Mechanics*, tr. T. J. McCormack (La Salle, IL: Open Court, 1960), pp. 287-88.

[10] J. Dewey, *The Quest for Certainty* (1929; New York: Capricorn Books, 1960), pp. 295-96.

[11] J. Dewey, "On the Nature and the Limitations of Cosmical Inquiries," *Scientific Monthly* 37 (1933), pp. 386-87.

[12] W. P. Bridgman, "The New Vision of Science," *Harper's Magazine* 158 (1929), p. 450.

[13] B. Russell, *The Scientific Outlook* (New York: Norton, 1931), p. 95.

[14] B. Russell, *Mysticism and Logic* (1917; Garden City, N.Y.: Doubleday, 1957), p. 45.

[15] Ibid., p. 94. In doing so, Bertrand Russell declared much of philosophy too to be a relic that can readily be discarded as he added that "the oneness of the world is an almost undiscussed postulate of most metaphysics" (ibid., p. 95).

[16] See *The Existence of God*, ed. J. Hick (New York: Macmillan Co., 1964), p. 174.

[17] W. James, *Some Problems of Philosophy* (New York: Longmans, Green and Co., 1911) p. 124.

[18] W. James, *A Pluralistic Universe* (New York: Longmans, Green and Co., 1909).

[19] W. James, *Pragmatism* (New York: Longmans, Green and Co., 1907).

[20] See ch. 4, "Champion of the Universe," in my book, *Chesterton, A Seer of Science* (Urbana, IL.: University of Illinois Press, 1986).
[21] W. James, *Pragmatism*, p. 3.
[22] James did not give a reference. The quote is in p. 9 of *Heretics* (London: John Lane 1905).
[23] W. James, *A Pluralistic Universe* (New York: Longmans, Green and Co., 1909).
[24] Ibid., p. 322.
[25] Ibid., p. 287.
[26] Ibid., p. 323.
[27] See for details ch. 6, "Loaded Dice," in my *God and the Cosmologists* (Edinburgh: Scottish Academic Press, 1989).
[28] W. James, *A Pluralistic Universe*, p. 325.
[29] Ibid., p. 328.
[30] Ibid., pp. 329 and 331.
[31] Ibid., p. 317.
[32] See my *Planets and Planetarians: A History of Theories of the Origin of Planetary Systems* (Edinburgh: Scottish Academic Press, 1978), pp. 52-53.
[33] B. Spinoza, *Ethics*, Part I, prop. xvi, in *The Works of Spinoza*, tr. R. H. M. Lewes (1883; New York: Dover, 1955), vol. 1, p. 59.
[34] Ibid., p. 60. In a letter to Oldenburg, Spinoza categorically stated: "As the nature of the universe is not limited, like the nature of blood, but is absolutely infinite, its parts are by this nature of infinite power infinitely modified and compelled to undergo infinite variations" (ibid., p. 292). Admirers of Spinoza, who love to quote this statement of his as a model of unparalleled insight, invariably fail to recall Spinoza's inability to reply to Tschirnhausen's question as to how the infinity, as conceived by Spinoza, could issue in particular beings. Only by ignoring these self-destructive facets of Spinoza's thinking can one set him up, as did Harry A. Wolfson, as the one who pulled down the triple scriptural philosophy (Jewish, Christian and Muslim), based on creation properly so-called, which starts always from particular concrete things. (*Religious Philosophy* [Cambridge: Harvard University Press, 1961], Preface). To see that philosophy as lying in ruins is as much a misconception about concrete history as is Wolfson's setting up of Philo as the founder of all medieval philosophy.
[35] According to H. W. N. Joseph, *Lectures on the Philosophy of Leibniz* (Oxford: Clarendon Press, 1949, p. 13), the two discussed Spinoza's *Ethics* and Descartes' theory of motion. Leibniz tried in vain to interest

Spinoza in his recasting of all human discourse in terms of the binary number system.

[36] "Absolutely speaking, it appears that God *can* make the material universe *finite* in extension; but the contrary appears more agreeable to his wisdom." See *Leibniz Selections*, ed. P. P. Wiener (New York: Charles Scribner's Sons, 1951), p. 246.

[37] Ibid., p. 543.

[38] *Monadology*, ibid., p. 544.

[39] *Encyclopédie*, vol. VIII (1765), p. 702.

[40] Originally published in 1855. Quotations are from the second edition of its English translation, *Force and Matter,* by F. Collingwood (London: Trübner and Co., 1870, pp. 13 and 47. The German original, translated into a dozen languages, had gone through nineteen editions before its author died in 1899.

[41] E. Haeckel, *The Riddle of the Universe,* tr. J. McCabe (New York: Harper and Brothers, 1900). p. 242. There Haeckel also categorically asserts the eternity of the universe and its transformation into all conceivable forms.

[42] The most succinct exposure of the internal contradiction of Spencerian cosmogenesis was given by H. G. Wells: "He [Spencer] believed that individuality (heterogeneity) was and is an evolutionary product from an original homogeneity, begotten by folding and multiplying and dividing and twisting it, and still fundamentally *it.*" *First and Last Things: Confession of Faith and Rule of Life* (London: Watts, 1929), p. 30.

[43] See the chapter, "The Angular Barrier," in my *Planets and Planetarians*, pp. 155-85.

[44] A. N. Whitehead, *Process and Reality: An Essay in Cosmology* (New York: Macmillan, 1929), p. 169. A curious assertion from one who about the same time spoke of Darwinists as the ones who "devote themselves to the purpose of proving that there is no purpose" and thus "constitute an interesting subject for study" (*The Function of Reason* [Princeton: Princeton University Press, 1929], p. 12). For if the universe contains no stable order, it is not possible to ascribe to it, without disregard for logic, an evolving order so that the result may differ from a random sequence of states that cannot even be properly called "states," that is, configurations with a recognizable measure of stability so that rational judgment may be made about them.

[45] H. Bergson, *Creative Evolution*, authorized translation by A. Mitchell (New York: The Modern Library, 1944), pp. 262-63.

[46] H. Bergson, *The Two Sources of Morality and Religion*, tr. R. A. Audra and C. Brereton (1935; Garden City, N.Y.: Doubleday, 1954), p. 317.
[47] K. R. Popper, *The Open Universe and Its Enemies* (5th rev. ed.: Princeton: Princeton University Press, 1966), vol. 1, pp. 202-03. Of course, as a "scientific" philosopher Popper tried to make it appear that he meant something very different from Bergson's vitalism.
[48] K. R. Popper, *The Open Universe: An Argument for Indeterminism* (Totowa, N. J.: Rowman and Littlefield, 1982), p. 143.
[49] K. R. Popper, *Conjectures and Refutations* (New York: Harper Torchbooks, 1968), p. 136.
[50] *The Open Universe*, p. 130.
[51] E. Mach, *The Analysis of Sensations and the Relation of the Physical to the Psychical*, tr. C. M. Williams (New York: Dover Publications, 1959), p. 243. The fact that Mach recognized the basic identity of his epistemological views with those of R. Avenarius, professor of philosophy in Zurich, gives special significance to the warning which Avenarius gave at the end of the Introduction to the second, posthumous edition of his *Der menschliche Weltbegriff* (Leipzig: O. R. Reisland, 1905), his exposition of a sensationist system which he called empirio-criticism. The warning meant to call in doubt the major difference between the 'universe' of one's sensory experiences and that given universe to which science lays claim. "On what ground, at any rate," Avenarius asked mockingly, can that difference be argued? (p. 3.) Already the reader of the Preface of the first edition could not be mistaken about the sad prospects which Avenarius' sensationism implied for a science of the universe. There he insisted that the universe could embody consistency only inasmuch as it was the universe of subjective sensations. Anything beyond was something "that could only mislead anyone, the more resolutely one tried to grasp it in its totality" (ibid., p. xii).
[52] R. Carnap, *Der logische Aufbau der Welt* (Berlin-Schlachtensee: Weltkreisverlag, 1928).
[53] H. Reichenbach, *Atom and Cosmos: The World of Modern Physics*, tr. A. S. Allen (London: George Allen & Unwin, 1932).
[54] H. Reichenbach, *The Rise of Scientific Philosophy* (Berkeley: University of California Press, 1951), p. 268.
[55] O. Neurath, "Sociology and Positivism," in A. J. Ayer (ed.), *Logical Positivism* (New York: The Free Press, 1959), p. 282.
[56] (New York: Holt, Rinehart and Winston, 1973). The book is the text of Ayer's Gifford Lectures, given at the University of St. Andrews, 1972-73.

[57] B. Magee, *Men of Ideas: Some Creators of Contemporary Philosophy* (London: British Broadcasting Corporation, 1978), p. 131.

[58] For Ayer's debunking of natural theology, see ch. 10, "The Claims of Theology," of his Gifford Lectures, quoted in note 56 above.

[59] J. A. Benardete, *Infinity: An Essay in Metaphysics* (Oxford: Clarendon Press, 1978), p. 103.

[60] The chief of them being E. Cassirer with his *Substance and Function and Einstein's Theory of Relativity* (authorized translation by W. C. Swabey and M. C. Swabey; 1923; New York: Dover Publications, 1953). That its index does not include the words—cosmos, cosmology, universe, world, and not even world-line—speaks louder than words.

[61] M. K. Munitz, *The Mystery of Existence: An Essay in Philosophical Cosmology* (New York: Dell, 1965). This work supersedes Munitz's earlier articles on the subject, published in *The Monist* and the *British Journal for the Philosophy of Science.*

[62] Ibid., pp. 58-61.

[63] Ibid., p. 63.

[64] Ibid., p. 66.

[65] See his *The Question of Reality* (Princeton: Princeton University Press, 1990) where he surveys modern scientific cosmology.

[66] Ibid., p. 68.

[67] Ibid.

[68] For texts of Athanasius, see my essay, "Christology and the Birth of Science," *Asbury Theological Journal* 45 (Fall 1990), pp. 61-72.

[69] M. Munitz, *The Mystery of Existence*, p. 68. Kant's critique of the idea of the universe occupies an all too visible place, a full chapter, in Munitz's *The Question of Reality*, pp. 65-79. There he builds on that critique by borrowing heavily from Wittgenstein's reduction of reality to words. Needless to say, Munitz does not carry this process of arguing to its full logic. Had he done so, he should have claimed that his book, or any book for that matter, exists only inasmuch as it is read aloud, and to the reader himself. Neither in *The Question of Reality*, nor in his earlier work, *Cosmic Understanding: Philosophy and Science of the Universe* (Princeton: Princeton University Press, 1986), did Munitz add anything essentially new to his arguments in *The Mystery of Existence.*

[70] A. Schopenhauer, *The World as Will and Idea*, tr. R. B. Haldane and J. Kemp (3rd ed.; London: Kegan Paul, Trench, Trübner, 1896), vol. 2, pp. 132-33.

[71] M. Munitz, *Cosmic Understanding*, pp. 281-82.

5

Proofs and a Proof

Between rationalism and empiricism there is only one middle ground, denoted by various terms, some more counterproductive than others. Aristotelianism, Thomism, Scholasticism, metaphysical realism, moderate realism, and similar labels are, like all labels, apt to create miscomprehensions. Moreover, those who call themselves Thomists represent nowadays a motley lot indeed, if they were ever really unified.[1] Aristotelians, old and new, also form disparate classes.[2] In one respect all of them are one, at least in modern times. Apart from avoiding the question, "Is there a universe?," they also say little about that largest of all conceivable objects. A rather striking disparity when seen against the large number of textbooks on cosmology, a subject, and even a label, which Scholastic philosophers had held as their exclusive domain for over two centuries until they ceded it to scientists.

Neglect by Scholastic philosophers of the universe as such is all the more surprising because a very different pattern was set for them by Christian Wolff. Those aware of the prodigious (though just as prolix) productivity of Wolff could expect something on a grand scale after Wolff published, in 1728, his *Discursus praeliminaris de philosophia in genere*, or the program of his philosophical synthesis. There he declared that "there is also a general understanding of the world which explains those things which are common to the existing world and to any other possible world. That part of philosophy which develops these general and abstract notions I call transcendental or general cosmology." Then he defined general cosmology as "the science of the world in general," adding that "general cosmology has in the past been unknown to philosophers, even though they have occasionally treated the things which pertain to it."[3]

This was, of course, true only in the sense that during the previous two thousand years philosophers did not take up that science in the systematic manner in which Wolff did in 1731 in his *Cosmologia*. It was the first major philosophical book (as well as textbook) with that word in its title.[4] There the first section, on "the notion of the world or universe," begins with a chapter on "the things' connection and their giving rise to a universe." The chapter, very verbose as all too often is the case with Wolff's discourse, has other and more serious defects as well. Wolff merely wants to lead to a definition of the world as "the series of finite entities, both simultaneous and successive, that are connected with one another"(§48). But, since he had already defined all things that change as necessarily finite, and all finite things as necessarily changeable (§45), he should have rather said that the totality of changeable things must also be finite. Only in proving that the world is contingent does Wolff refer to infinite regress as a wholly insufficient explanation (§§91-93). Again, something much more demonstrative could have been stated about integers than that all essences are like rational integers (§60), in order to prove that the world is one.

The weakness of Wolff's philosophical speculations about the world or universe was mirrored in his scientific account of it which first appeared in 1723 and became quickly known as "German Physics."[5] It showed no appreciation of Newton even in its subsequent editions,[6] which was also true of Wolff's excursions into science in his Latin works. The same applies to Wolff's discourses on natural theology where he offered, in an almost obsessive manner, examples of teleological arrangements in every nook and cranny of the universe. There the world as a whole figured as an "image of God," a more metaphorical than metaphysical expression.[7]

From the viewpoint of science Wolff's dicta on the universe were even more desperately behind times around the mid-1750s when Alexander Gottlieb Baumgarten rescued Wolff's rationalism from the sudden eclipse it had suffered in the 1730s. Baumgarten made no effort whatever to recall in Part II (Cosmologia) of his *Metaphysica* Wolff's way of proving the existence of the universe. Rather, Baumgarten proved the possibility of the universe from its existence![8] One of the reasons for this slighting of the universe on Baumgarten's part was his preoccupation with aesthetics,[9] which suggested a trend to come.

Aesthetics, as a form of intuitionism, served in Kant's *Critique of Judgment* as the means of justifying references to the universe. Kant, however, cautioned against taking this avenue to the "universe" for an avenue to the Universe or an absolute totality of things. After allowing that the "universe" could be taken for an illustration of the "mathematically sublime," he recalled a principal contention of his in the *Critique of Pure Reason.* He insisted once more that "in the logical estimation of magnitude, the impossibility of ever arriving at absolute totality, by means of the progress of the measurement of things of the sensible world, was cognized as objective, i. e. as an impossibility of *thinking* the infinite as entirely given, and not as merely subjective or that there was only an incapacity to *grasp* it."[10]

This Kantian claim helped to bring about a thorough lack of interest in a philosophical approach to the universe itself within Protestant ambience.[11] The process had been prepared by the always ambivalent interest there in natural theology. Rather logical was the enthusiastic reception given in Protestant circles to Kant's endorsement, in his *Critique of Practical Reason,* of a "practical" approach to God, either through hearing the voice of one's moral conscience or through one's contemplation of the starry sky, a sort of aesthetics. A sharp light was shed on this by the famed mathematician, M. Cantor, when in a letter to Bertrand Russell he remarked on Kant's popularity among Protestants: "German Protestantism in his development to 'Liberalism' *needs* himself [itself] *a fundament* on which to build his [its] *seeming*-Christianity, so Kant or one of his successors are picked out, by the Protestant theologians of diverse schools, to be their *Atlas*. One hand washes the other, one depends on the other and *one has to fall with the other!*"[12]

Irony is not absent in the fate of the universe in cosmologies written by Catholic philosophers, including those who worked after Pope Leo XIII called, in 1879, for a reorientation toward the thought of Thomas Aquinas. The call came less than a decade after Vatican I which solemnly restated man's ability to recognize, from things visible, their Creator. While this was not an endorsement of the cosmological argument, its study could at least be seen as greatly encouraged. Such should seem all the more the case as the Council continued the opposition of Gregory XVI to the fideism championed by Louis Bautain and others a few decades earlier.[13]

The irony of the neglect by Scholastic philosophers of the universe becomes even more pronounced when seen against the credal statements of Christian faith. All those statements are almost entirely about the outstanding events of salvation history, but they all rest on the starting proposition about the creation of all by God. Wholly gratuitous as Redemption may be to Creation, the former is totally dependent on the latter. This is why all credal statements of Christianity (the only religion with credal statements) begin with a reference to the Father Almighty as the Creator of all. As to the credal phrase, "of all things visible and invisible," already present in the Nicene Creed, it has failed to become a standard phrase for the physical universe, perhaps because the term "invisible" could be taken for purely spiritual beings, such as angels.

Little is known about the reasons for the insertion, from the Niceno-Constantinopolitan Creed (381 A. D.) on, into the Creed of the phrase, "heaven and earth," the classic Old-Testament idiom for the totality of things.[14] The phrase, which had to appear primitive already within the context of Ptolemaic astronomy, has a saving grace in post-Copernican cosmological perspectives only through its archaic flavor. At any rate, none of the Creeds, and certainly not the Niceno-Constantinopolitan Creed, which is the most universally used Creed throughout Christendom, proposes the existence of the universe as an object of faith. The universe is rather part of the credal statements as that very object which is so evident as to prompt one to acknowledge the Maker of All, or the Almighty Father.

The demise of the geocentric universe exposed the futility of taking the universe for something empirically observable. The cosmological argument should seem therefore to call for a proof that there is a cosmos or universe. Yet the Scholastic tradition, as represented by its textbooks on cosmology, contains no explicit argument that there is a strict totality of things or universe. In fact, only a minority of these textbooks on cosmology contain a chapter on the cosmos, though never the argument in question. Their authors assume the existence of the universe when, for instance, they discuss the question whether there is only one universe or there are many universes.[15] Their failure to confront the challenge posed by Kant and his influence is especially revealing when on occasion they take up the philosophy of the universe.[16] The penetration of Kant into Catholic thinking through Transcendental Thomism explains, of course, the short shrift given to the universe in cosmologies written by adherents to that new-fangled Thomism.[17] That the universe was treated like a foster child even by some of those who supposedly had the greatest stake in it, should seem baffling to say the least. They still have to learn that salvation history makes no sense if there is no creation properly so called, that is, a creation of all, or the Universe, out of nothing.[18]

As was already noted, the question whether the universe is one is not entirely the same as the question whether there is a universe. Aristotle argues the first question in some detail and takes for granted the affirmative answer to the second. In that latter respect he may have been motivated by his belief that in the sphere of the fixed stars one could see the very confines of the universe. His strongest grounds for taking the universe for a reality were, however, of an a priori character mixed with teleological considerations. Suffice to recall the breathtakingly sweeping chapter in *On the Heavens*,[19] where he derives in a short page all the characteristics which the universe must necessarily have.

Aristotle did not elaborate on his a posteriori approach, or his statement that "since the world (*ouranos*) is perceptible, it must be reckoned a particular, for as we have seen, everything that is perceptible, is in matter."[20] This potentially valuable starting point in proving that there is a universe does not reappear in his paraphrase of the most comprehensive among the three meanings of the word *ouranos* (world): "the whole which is enclosed by the outermost circumference must of necessity be composed of the whole sum of natural perceptible body." And a page or so later: "The world in its entirety is made up of the whole sum of available matter (for the matter appropriate to it, as we saw, is natural perceptible body), and we may conclude that there is not now a plurality of worlds, nor has there been, nor could there be. This world is one, solitary, and complete."[21]

The meaning Aristotle intended to convey with the word "complete" is that the universe is necessarily what it is and cannot be anything else. Aristotle did not even doubt that even the size of the universe had to be what it was. His interest in quantities was philosophical, though it reveals, for instance, considerable familiarity with the works of Eudoxus.[22] His chief concern was the one bequeathed by Socrates on Plato, Aristotle's own teacher. Socrates took the view, and in a most memorable context, that his own personal sense of purpose could not be justified unless he could show that all material bodies moved for a purpose.[23]

This is the origin, an epistemologically most mistaken origin, of the Aristotelian doctrine of cosmic coherence in terms of natural places and natural motions. The strict unity of things is assured by their natural purpose. It is on such grounds that Aristotle rejects Democritus' speculations about a very large, in fact, an infinite number of worlds, and not with a reference to the impossibility of an actually realized infinite quantity, although he upholds that impossibility. Were large masses (worlds) composed of stone or water outside the spherical boundaries of our universe, they would, Aristotle argues, rush towards that center in our universe which is the target of the motion of all earth and water. From the same presuppositions it also follows for Aristotle that the only world cannot be infinite, because in such a world there would be movement with infinite speed. This means for him not so much a physical impossibility as a sort of ethical incongruity: ordinary bodies cannot reach immediately and permanently their natural place which is the fulfillment of their nature or purpose.

As most other Greek philosophers of old, Aristotle too was (*pace* Brentano and a few others) a pantheist and, therefore, an eternalist. As such he anticipated the attitude of all pantheists, and of their variety known as materialists, toward the universe. They all seem to know, as if by an innate knowledge, that there is a universe and that it is the only truly existing entity. They were also anticipated by Aristotle as he claimed that the actual single universe comprises every time element which are infinite in number because the universe is "immortal and divine."[24] Such a view of the universe entails for Aristotle that the Prime Mover is not a personal God, a Creator. In the very same context, in the last book of his

Metaphysics, where he speaks most thematically of the Prime Mover, Aristotle states that the universe is like a house without a master, or like an army without a commander.

Unlike the Averroist or strictly Aristotelian interpreters of Aristotle, who, like their master, did not believe in a Creator and treated the universe as he did, Thomas Aquinas was a theologian who took the reality of the Creator for the principal perspective in dealing with any question of theology. Thomas therefore proved not in a philosophical but in a theological way that there had to be a strict totality of things. As a perfectly rational being the Creator could only create things that were fully coherent with one another. Since God creates things for a reason, the world is one, "because all things must be arranged in one order, and to one end."[25] Like all his contemporaries, Thomas too believed that he could see the confines of the universe by looking at the sky. Possibly, for this reason, Thomas did not explore the purely philosophical way of proving that there was a strict totality of things. Another reason was that Thomas too was no exception to the rule that the intellectual horizon of any philosopher is limited by the horizon of his time. Thus Thomas' philosophical discussion that the universe is one is largely definitional and in this respect too along the lines of Aristotle's dicta.

That there is a philosophical way of proving that there is a universe, was merely hinted by Thomas as he gave his Five Ways of proving the existence of God. There, as is well known, he rejects regress to infinity (taken in a metaphysical sense) as a mere postponement of the answer about the sufficient reason for the reality, or the universe, on hand. In any of those five ways the starting point is the material object or thing insofar as its existence is registered or recognized. On a cursory look this may appear a purely empirical starting point, but for Thomas it includes an emphasis on the existence of the thing in question and on its being perceived in its rationality, that is, representing something general in its individuality. These features, which are missing in typical empiricist, let alone sensationist, epistemologies, are the hallmarks of Aristotelian epistemology, features further enhanced in its Thomistic rendition.

Such a starting point was not spelled out by Thomas with the intention of forging a proof of the existence of the universe. It may, however, be worth considering whether once an appropriate starting point had been chosen from Thomas' writings, one may put together from his various utterances on the universe, which are startlingly numerous,[26] a proof of its existence. The choice of a philosophical starting point in Thomas' philosophical dicta is all the more a delicate question as only a few and relatively short writings of his qualify for systematic philosophical discourse. His fondness for systematizing always related to theology. This is why in his most systematic theological discourse, the *Summa theologica,* Thomas begins with God, the One, and has always God in focus even when he talks about anybody or anything else. Since for him theology is a rational discourse about whatever is known about God, either through nature or through revelation, he brings in philosophical considerations at every step though not for their own sake.

Among such considerations or philosophical gems is a remark of his in question 11 of Part I of the *Summa theologica,* a question in three articles on the "Unity of God." The first article is an answer to the question "whether one adds anything to being." After listing the possible objections,

Thomas gives his succinct reply and, in this case, he does so by quoting a dictum from a sixth-century book on *Divine Names* which in his day was still believed to have been written by Dionysius the Areopagite, a disciple of Paul: "Nothing which exists is not in some way one, which would be false if *one* were an addition to *being,* in the sense of limiting it" (emphasis added).[27] What Thomas here states is that the numerical oneness of an entity is a sequel, in the ontological order, to its being.

From this it follows, and in the second article Thomas spells this out explicitly, that multitude as such cannot be known. The relevance of such a thesis to our task should be obvious through a mere recall of the multitude of objects that constitute the scientific or telescopic universe. But, before dealing with that relevance, let us hear Thomas for a little longer. According to him multitude is understandable insofar as it implies oneness. But then what is that division that produces the many from the one? For modern scientific minds division is a mere operational procedure with quantitities. Thomas sees division in a rather different light. According to him division is understood only "from the very negation of being: so what first comes to the mind is being; secondly, that this being is not that being, and thus we apprehend division as a consequence; thirdly, comes the notion of one; fourthly, the notion of multitude."[28]

This sequence may sound very abstract, eminently reasonable though it is. Luckily, concrete illustrations are legion. Every noun is an example. Modern astronomical parlance provides many instances, such as clusters, galaxies, supergalaxies and so forth. What we really see are their components. And even when we focus on those components, mostly stars, we should not forget that the word "star" is a means whereby a multitude of atoms is reduced to some unity as it is being understood. Every noun, whether used in astronomy or elsewhere, has this property of being a universal. The very closeness of "universal" to "universe" may suggest the correctness of basing on this doctrine about the universals the validity of knowing whether there is a universe.

Not that Thomas reaches the universe, this most inclusive of all nouns or universals, along that way. But his dicta on the universals provide helpful directives. He urges a close look at the relation of the particular to the universal in terms of which the particular or individual is understood. Man as such does not concretely exist. Only the individual man, denoted by the phrase, this or that man, does exist: "Hence the thing which is a man has something more in it than has humanity. Consequently, humanity and a man are not wholly identical; but humanity is taken to be the formal part of a man, because the principles whereby a thing is defined are regarded as the formal constituent in regard to the individualising matter."[29] Such is Thomas' basic dictum about the relation of the individual to the universal. That Thomas' penetrating epistemological observation comes in the context of his discussion of the simplicity of God proves that it is not necessarily a waste of time for philosophers to read theology. It can in fact be very useful for physicists and, especially, for cosmologists who try to face up to the question how they know that there is a universe which is more than the observed or operationally observable sum of celestial bodies.

Clearly, for Thomas the universe cannot be an individual, an entity with consciousness, or even endowed with that life which we observe in any animal or plant. Between the geocentrism of the Socratic or Greek tradition and the geocentrism of Thomas there is indeed a world of difference, whatever their apparent identity. In the universe of the pagan, or at best, pantheistic Greeks, the unity of all was derived from a quasi-conscious volition involving all. There even the most distinct units, as embodied in personal consciousness, were to be absorbed into a single cosmic intellect that lacked personal identity. Thomas' universe, in which individual consciousness has an eternal permanence, is, for that reason too, a genuine multitude of things and therefore a genuinely modern or scientific universe.

How then does Thomas know that the multitude of things is a unity, a true universe? He could have taken the easy way out by saying that he sees the universe through seeing its spherical boundary. Had he done so, his dicta would be useless for our purposes because relativistic cosmology implies the impossibility of observing the boundaries of the universe. Those boundaries correspond to the largest net of the curvature of the gravitational field, a net giving the permissible paths of motion. Since there is no way of going beyond that net, modern scientific cosmology implies the impossibility for any bodily observer to move beyond the universe, even if that impossibility were not implied in the very notion of the universe itself.

Thomas argues the unity in the multitude on a ground which is scientific in such a way as to allow a genuine philosophical recasting. The way is the relatedness of scientific law to order on the ontological level. Otherwise, scientific laws cannot be about reality. For Thomas the universe is a multitude of individual things, but of things in each of which there is an order and which can interact with one another on the basis of the same orderliness. Things, he says, have an order to each other, and "hence it must be that all things should belong to one world."[30]

Since things for Thomas are always creations and God is always the Creator, Thomas speaks with no trace of hesitation about the orderly interconnectedness of things (a most scientific perspective). Therefore, apart from the basic epistemological argument that to know is to know the object insofar as some wholeness or universality is revealed in it, Thomas also argues that the unity of things is a consequence of the rationality of God's creative act. Since God, who is supreme rationality, creates things with a reason, the world is one, "because all things must be arranged in one order, and to one end."[31]

This rational and orderly action of God is a decisive consideration for Thomas when he discusses the question whether an infinite multitude can exist, in article 4 of question 7 of Part I of the *Summa theologica.* In Thomas' words, "everything created is comprehended under some clear intention of the Creator; for no agent acts aimlessly. Hence everything created must be comprehended in a certain number. Therefore it is impossible for an actually infinite multitude to exist, even accidentally,"[32] that is, when an another infinite magnitude need not be postulated for the production of the quantity in question. Thomas admits only the infinity which is on hand in the continuation, to no end, of the division of a given magnitude.

The strength of Thomas' conviction about the impossibility of an actually realized infinite creature or material body is further revealed by his insistence in the *Summa* that even a pure number cannot be actually infinite.[33] Thomas therefore provided such parts of a proof of the existence of the universe as the invariable bearing of human knowledge on wholeness and of the impossibility that the most encompassing of material wholeness should lack specific finiteness or limitedness. Still Thomas did not turn these considerations into a proof of the existence of the universe as a strict totality of things. Of course, as has been stated above, no philosopher can be expected to discuss something which is no problem to any of his contemporaries. The existence of the universe became a pressing problem only with the coming of the telescope which destroyed a universe enclosed within the sphere of the fixed stars. A big scientific gain as this could be, it also led to an apparent loss of cosmic cohesion: an allegedly infinite Newtonian universe could not be seen to cohere gravitationally.[34] Kant's pseudo-Newtonian attacks on the universe were to come in due course. But precisely because of the distrust which Kant produced in the modern mind about a genuinely metaphysical epistemology, the effectiveness of what can be gathered from Thomas for a Thomist proof that there is a universe, should seem meager indeed.

If the modern mind still retains some metaphysical sensitivity, it is heavily tainted with Kantian a priorism. Therefore a demonstration of the existence of the universe from a consideration of universals will be easily taken for a mere version of the same a priorism with which Kant handled the categories. After all, even many latter-day Thomists are still to take seriously Thomas' emphatic declaration that the categories of the mind are derived from sensory experience.[35] Otherwise it would not have occurred to them to propose and work out that miscegenation of Aquinas and Kant, known as Transcendental Thomism, better called Aquikantism.

Apart from its metaphysical preconceptions, if any, the modern mind is all too heavily tuned to science and, as a result, is ready to grow sceptical to the point of letting that greatest prize, the universe, escape its hold. That mind may be more receptive to an argument which is still Thomist in the sense that its starting point is concrete material reality. It is as concrete as the stone kicked by Samuel Johnson; it is not demonstrable in any other way; it fails to impress only those (unfortunately many) who are so obsessed by ideas, by critical knowledge, and the like taken for a starting point in reasoning that in the long run even stones degenerate into mere ideas in their ever more convoluted utterances.

It should seem obvious that material reality can be such only if it has quantitative, that is, countable properties. Matter which is not countable may be of interest to very speculative philosophers but it is useless for scientists. Scientists in recent years have come up with exceedingly esoteric forms of matter and fields of force that may appear sheer abstractions when contrasted with ordinary matter. Yet even those esoteric forms must have countable properties in order to be scientifically useful. The interaction of one piece of matter with another, and of all pieces of matter with one another, is meaningful for science as long as it can be measured, that is, counted.

Counting is, however, always done in terms of integers. In other words, matter is to be taken to exist in a way which can also be comprehended in terms of integers, the basis of all counting. Whatever else Krönecker meant by saying that "God made the integers and everything else is the work of man,"[36] he certainly wished to express the fundamental role which integers play in mathematics. The coherence of all material things can therefore be seen as a consequence of the universal applicability of integers wherever matter exists. The specific kind of coherence is not, however, revealed by the countability of matter but must be found out empirically. Modern physicists have indeed expressed wonderment time and again about the fact that of the great many mathematical systems only one seems to fit the physical universe.[37]

The belief that this universality is valid throughout cosmic spaces received a palpable endorsement in the code engraved on Pioneer X, a code aimed at alerting denizens of extraterrestrial civilizations to our existence on earth.[38] To be sure, Sagan and other champions of extraterrestrial intelligence were not willing (or able) to consider the metaphysical implications of their procedure. For if life is an accidental occurrence and its evolutionary development is purely random, it is difficult to see why that most unlikely evolutionary product, the thinking mind, would occur in the same form everywhere else. Sagan and others would not have gone even as far as to temper their enthusiasm for evolutionary randomness by pondering a most considered remark of no less a mathematician than John von Neumann: "Anyone who considers arithmetical methods of producing random digits is, of course, in the state of sin."[39] Of course, in order to appreciate the weight of this statement one has to keep in mind the absolutely fundamental role which integers play in arithmetic and, therefore, in all mathematics, including probability theories of all sorts. Even the most "random" among them cannot operate without a consistent application of integers.

Munitz and others, who do not find it demonstrated that the material universe, taken in its absolute totality, is coherent in all its parts, are still to demonstrate that matter can exist without being subject to that counting that can, in ultimate analysis, only be made in terms of integers. No numbers, however "irrational" or "complex" they may be, can be defined with no reference to integers. As to fractions, the very word is meaningless unless there is a unit that can be fractioned, that is, broken into parts. No integers can, however, be actually infinite as such an infinity is a denial of its actual countability. Consequently, countable or measurable matter too has to have its own boundaries.

This boundary is not necessarily the horizon as specified in reputable forms of scientific cosmology. Basic reflections on the history of science should be enough of a warning against taking this or that phase of science for the last word. At no point can the sum of empirical evidence be taken to be absolutely complete. At no point can one exclude the possibility of some future findings that demand major revisions of a theory, however successful.[40] Apart from this there is a theoretical ground against taking this or that form of scientific cosmology for a necessarily true account of the universe. Any such form has to be highly mathematical and therefore cannot have its proof of consistency within itself as long as Gödel's incompleteness theorems remain valid.[41]

A proof of the universe which takes its start from a matter which is countable assures to the universe a boundless character very different in meaning from the one usually encountered in scientific cosmology. There it usually means the impossibility of reaching physically the boundary of the universe in the sense of going through it. Therefore even the spherical and static Einsteinian universe can be called boundless. If its four-dimensionality is illustrated in a three-dimensional spherical surface, there any ordinary three-dimensional entity (reduced to a two-dimensional extension, such as an infinitesimally thin coin) can slide forever without coming to a boundary. The boundless or open universe implied in a model of the universe which is forever expanding is not what is meant here either.

In the proof in question one is faced with an openness or boundlessness of the universe because the integers which assure countability do not by themselves reveal the kind of mathematical system which is embodied in the structure of matter. The point is missed in the notable but partly misleading reference to "the unreasonable effectiveness of mathematics" in physics.[42] One should rather speak of the elusive identity of the exceedingly effective and even heuristic countability which matter possesses. While that countability can support a proof that there is a material totality, always limited and therefore bounded, which is the Universe, it also assures a boundless character to its scientific investigation. This is a consideration which is not part of the proof itself while it may add to its convincing value, a difference to be explored in the next chapter.

[1] See, for instance, Helen J. John, *The Thomist Spectrum* (New York: Fordham University Press, 1966).

[2] As can be seen from paging through any few volumes of the *Proceedings* of the Aristotelian Society (London).

[3] C. Wolff, *Preliminary Discourse on Philosophy in General*, tr. with an introduction and notes by R. J. Blackwell (Indianapolis: The Bobbs-Merrill Co., 1963), p. 41.

[4] *Cosmologia generalis, methodo scientifica pertractata, qua ad solidam, inprimis Dei atque naturae, cognitionem via sternitur* (1731; editio nova priori emendatior; Francofurti et Lipsiae: Officina libreria Rengeriana, 1737). For a list of earlier and rather insignificant books with the word *cosmologia*

in their titles, see note 1 (p. 275) in my translation of Kant's *Universal Natural History and Theory of the Heavens* (Edinburgh: Scottish Academic Press, 1980). Wolff's *Cosmologia* was the second of the five great works in which he set forth in Latin his speculative philosophy between 1730 and 1737. Wolff's Latin work was an elaboration on a vast scale of his "German Philosophy" which he set forth in eight volumes between 1719 and 1725, the title of each of which began with the words, "Vernünftige Gedancken." Of these the second, *Vernünftige Gedancken von Gott, der Welt und der Seele des Menschen, auch allen Dingen überhaupt* (1720), was his "German Metaphysics." It dealt, in its ch. 4, with the notion of the world, of bodies, and of the perfection of the world, a sort of cosmology. The importance of cosmology in the Wolffian system was revealed to Wolff by his foremost champion, Ludwig Philipp Thümmig, who produced in 1725-26 his two-volume *Institutiones philosophiae wolffianae*, a fact which Wolff duly acknowledged. For details, see M. Wundt, *Die deutsche Schulphilosophie im Zeitalter der Aufklärung* (Tubingen: J. C. B. Mohr, 1945), pp. 191 and 212.

[5] *Vernünftige Gedancken von den Würkungen der Natur* (Halle im Magdeburg: Rengerische Buchhandlung, 1723).

[6] Of which four appeared prior to Wolff's death in 1754.

[7] *Theologia naturalis*, published in two volumes in 1736 and 1737. It gave rise to books on the theology (or rather teleology) of stones, of water, of air, etc., an outcome which gave plenty of ammunition to critics of natural theology.

[8] See the photocopy reprint of the 7th edition (1779) by Photographischer Betrieb, Darmstadt, 1963, p. 111. Baumgarten's disinterest in cosmology is revealed by the list of his publications in *Gesamtverzeichnis des deutsch-sprachigen Schrifttums*, vol. 9 (München: K. G. Saur, 1979), pp. 215-17.

[9] A point rightly stressed in G. Tonelli's article on Baumgarten in *The Encyclopedia of Philosophy* (New York: Macmillan, 1967), vol. 1, pp. 256-57.

[10] I. Kant, *Critique of Judgment*, tr. J. H. Bernard (New York: Hafner Publishing Co., 1951), p. 98.

[11] The process is well illustrated in Wundt's work quoted in note 4 which should be studied in connection with his *Deutsche Schulmetaphysik des 17. Jahrhunderts* (Tubingen: J. C. B. Mohr, 1939). There one can see many instances of early Protestant scholastic interest in natural theology, largely through the influence of the works of the Jesuits, P. Fonseca and F. Suarez. The almost total neglect of natural theology by Protestant

divines during the 19th century is portrayed dramatically, though with careless documentation, in K. Barth's *Protestant Theology in the 19th Century* (London: SCM Press, 1972). Rather revealing is the absence of any discussion of cosmology in J. Dillenberger's *Protestant Thought and Natural Science: A Historical Interpretation* (Garden City: N.Y.: Doubleday, 1960).

[12] *The Autobiography of Bertrand Russell. Volume I. 1872-1914* (London: George Allen and Unwin, 1967), p. 219. The letter is dated September 19, 1911.

[13] A remark pointedly made in G. Boas' article on Bautain in *The Encylopedia of Philosophy*, vol. 1, p. 257.

[14] This idiom is one of those "additional words and clauses" in the Niceno-Constantinopolitan Creed which were characterized as of "little or no significance" by J. N. D. Kelly in his *Early Christian Creeds* (London: Longmans, 1950), p. 303.

[15] It is in discussing the question whether the universe is one that authors of such textbooks come closest to our question, without, however, raising it. Examples of this, to mention only 20th-century works, are J. J. Urráburu *Cosmologia* (Madrid: Gabriel López, 1903); *A Manual of Modern Scholastic Philosophy,* written under the direction of Cardinal Mercier by professors of the Institut Supérieur de Philosophie of Louvain, in two volumes (3rd English ed.; London: Routledge and Kegan Paul, 1960), which begins with a section on cosmology; J. Donat, *Cosmologia* (5th ed.: Innsbruck: F. Rauch, 1924); G. Esser, *Cosmologia* (Techny, Ill: Domus Missionis ad St. Mariam, 1939); P. Hoenen, *Cosmologia* (5th ed.: Rome: Gregorian University, 1956); F. Selvaggi, *Cosmologia* (2d ed.: Rome: Gregorian University, 1962); J.-M. Aubert, *Philosophie de la nature* (Paris: Beauchesne, 1965). The complete or almost complete absence of references to the universe is exemplified by J.-M. Dario, *Praelectiones cosmologiae* (Paris: Beauchesne, 1923); D. Nys, *Cosmologie ou étude philosophique du monde inorganique* (4th ed.: Louvain: Em. Warny, 1928); F. Remer, *Cosmologia* (Roma: Universitas Gregoriana, 1933); M. Fatta, *Cosmologia* (Milano: Vita e Pensiero, 1936); F. Saintonge, *Summa cosmologiae seu philosophiae naturalis generalis* (Montréal: Imprimerie du Messager, 1941); F. Renoirte, *Cosmology: Elements of a Critique of the Sciences and of Cosmology,* tr. J. F. Coffey (New York: Joseph F. Wagner, 1950); R. Masi, *Cosmologia* (Rome: Desclée, 1961). The matter would deserve a detailed study.

[16] Thus, for instance, J.-M. Aubert, *Philosophie de la nature: Propédeutique à la vision chrétienne du monde* (Paris: Beauchesne, 1965) and F. Selvaggi, *Filosofia del mondo: Cosmologia filosofica* (Roma: Università Gregoriana Editrice, 1985).

[17] A glaring case is *Cosmos* (New York: Fordham University Press, 1973) by R. J. Pendergast, who tellingly enough dedicated it to Karl Rahner and Teilhard de Chardin. The latter, it is appropriate to note, never considered in his voluble meditations and perorations on the universe the question of what were his grounds for taking it for a reality.

[18] As I argued in my Père Marquette Lecture, *Universe and Creed* (Milwaukee: Marquette University Press, 1992).

[19] *On the Heavens*, tr. W. K. C. Guthrie (Cambridge, Mass.: Loeb Classical Library, 1960), 286a-b.

[20] Ibid., 278a.

[21] Ibid., 278b-279a.

[22] As shown in great detail in Sir Thomas Heath's posthumous work, *Mathematics in Aristotle* (Oxford: Clarendon Press, 1949).

[23] A point which I often insisted upon, most recently in my essay, "Socrates or the Baby and the Bathwater," *Faith and Reason* 16 (Spring 1990), pp. 63-79.

[24] *On the Heavens*, 279a.

[25] *Summa theologica*, Part I, qu. 47, art. 3.

[26] As can easily be gathered from the almost three thousand entries under the words 'mundus' and 'universum' in volumes 14 and 22 of *Index thomisticus*, ed. R. Busa (Stuttgart/Bad Cannstatt: Frommann-Holzboog, 1975). Those entries reveal the richness of Thomas' reflections on the universe as portrayed in my essay, "Thomas and the Universe" (*The Thomist* 53 [1989], pp. 545-72), where, however, no effort was made to construct the proof which is under consideration here.

[27] *Summa theologica*, Part I, qu. 11. art. 4. This and subsequent quotations from the *Summa* are from its English translation in three volumes, literally translated by the Fathers of the English Dominican Province (New York: Benziger Brothers, 1938), p. 46.

[28] Ibid., p. 47.

[29] Ibid., Part I, qu. 3, art. 3 (p. 16).

[30] Ibid., Part I, qu. 47, art. 3 (p. 247).

[31] Ibid., p. 248. In the same breath Thomas notes, with a reference to Democritus, that "those only can assert that many worlds exist who do not acknowledge any ordaining wisdom, but rather believe in chance."

[32] Ibid., Part I, qu. 7, art. 4 (p. 33).

[33] Ibid., p. 33. "Every kind of multitude must belong to a species of multitude. Now the species of multitude are to be reckoned by the species of numbers. But no species of number is infinite; for every number is multitude measured by one." These passages are, however, to be balanced with Thomas' assertion that while even God "cannot make a thing to be not made because this would imply that two contradictories are true at the same time" (qu. 7, art. 2, p. 32), there is no contradiction in the notion of what is relatively infinite. As an example he mentions the potentiality of matter to take on an infinite number of shapes. He left without explanation a much quoted remark of his, which is found at the end of his public lecture, "On the Eternity of the World." There, in discussing the theological argument against infinity, namely, that the number of souls cannot be infinite, he concludes that "no demonstration has as yet been forthcoming that God cannot produce a multitude that is actually infinite." Quoted from *St Thomas Aquinas, Siger of Brabant, St Bonaventure, On the Eternity of the World*, translated with an introduction by C. Vollert, L. H. Kendzierski, and P. M. Byrne (Milwaukee, Wis.: Marquette University Press, 1964), p. 25. Perhaps this was on Thomas' part an aside, made in the heat of dispute, not characteristic of him.

[34] Part of the theme of ch. 1, "Universe Regained," in my *God and the Cosmologists* (Edinburgh: Scottish Academic Press; Washington, D.C.: Regnery Gateway, 1989).

[35] Clearly, if even Bonaventure's ideas about innate notions cannot be reconciled, as E. Gilson argued in his *The Philosophy of St. Bonaventure* (tr. I. Trethovan and F. J. Sheed; Paterson: St. Anthony Guild Press, 1965, p. 462), with Kant's understanding of the categories, this should seem even more true about a reconciliation of Aquinas with Kant. Whereas for Aquinas even the categories of the mind are the fruit of sensory perception, Kant would still have to argue, and in vain, that the *Critique* does not impose a consideration of Euclidean space perception as necessary in an a priori manner.

[36] Krönecker made this statement in 1866 at the Naturforscher Versammlung in Berlin.

[37] Thus, for instance, E. P. Wigner, "The Unreasonable Effectiveness of Mathematics in the Natural Sciences," *Communications in Pure and Applied Mathematics* 13 (1960), pp. 1-12.

[38] For a reproduction of that plaque, see Illustration XLI in my *Planets and Planetarians: A History of Theories of the Origin of Planetary Systems* (Edinburgh: Scottish Academic Press, 1978). Of course, instead of Arabic numerals, obviously arbitrary symbols even on the Earth, short lines

indicated the number of satellites around the various planets, as if lines were necessarily more universal symbols.

[39] Quoted on the very first page of D. E. Knuth, *The Art of Computer Programming. Volume 2, Seminumerical Algorithms* (Reading, Mass.: Addison-Wesley, 1969), p. 1.

[40] This may mean the finding of far more esoteric kinds of matter than the so-called "missing mass." Yet, already that mass or matter prompts some cosmologists to conjure up the vision of a universe "radically" different from the one known. Worse, this cavalier handling of what is radically different is labeled by them (see J. Trefil's approving report in the June 1993 issue of *Smithsonian Magazine)* as the ultimate step in unfolding the meaning of the Copernican revolution or principle. The fact is that Copernicus never played irresponsible games with human rationality in its relation to the universe. See my *The Purpose of it All* (Edinburgh: Scottish Academic Press, 1990), pp. 96-99.

[41] A point argued in detail in chapter 4, "Gödel's Shadow," in my *God and the Cosmologists*. It speaks louder than words that there is not a single reference to Gödel in S. Weinberg's *Dreams of a Final Theory* (New York: Pantheon Books, 1992).

[42] The title of E. P. Wigner's article, quoted in note 37 above.

6

Proof and Convincing

The difference between a proof and the conviction it ought to generate poses a special problem in an age conditioned to the proofs which exact science can deliver. About those proofs it is easily overlooked that all too often they amount to mere identity relations, dressed as they can be in the language of mathematical physics. Its equation signs readily convince because the two sides they connect contain no real novelty with respect to one another. Thus the conviction which those proofs or demonstrations generate comes forth almost automatically. Taken by themselves, they are of little help to convince one about the reality to which they supposedly refer. For no equation sign can be set up between the mind, swayed as it can be by the identity relations of logic or mathematics, and external reality.

That even the fact, let alone the nature, of external reality, however ordinary, cannot be proven by mere logic or mathematical formulas does not make one's immediate registering of external reality a less than fully rational process. To know the existence of things is in fact the very first step in reasoning. Any "critical" knowledge or philosophy which does not accept this will remain a mere criticism of criticism and not a criticism of the external reality one registers, and not even of one's registering it.

This should be acknowledged even by rigid idealists insofar as they want to communicate to others their claim that it is possible to start with mere ideas. Only by at least tacitly assuming the priority of external reality can they engage in a dispute with others about the proof of the Universe which takes its starting point from registering the existence of matter and goes on to stating that matter can be counted or else it is something else. This is, of course, but an aspect of the rule which should seem basic to all philosophy with a claim to a minimum of consistency: No form of philosophy, not even idealism (let alone its Kantian brand), can assert less than the means whereby it is communicated to others. Otherwise philosophy degenerates into a non-communicable love of wisdom, a sort of radical self-centeredness. Among those means is such an indispensable, material factor as the vibrations of the air, produced by flesh-and-blood tongues (and not merely by their critical ideas). Those air-waves must impinge on real ears in order to initiate an argument. The point can be made with even greater force about books as carriers of philosophical arguments. Unlike evanescent air-waves, books remain tangibly present even after the arguments fell silent together with the disputants.

Such considerations should make *prima facie* suspect two kinds of discourse about the Universe. One is riveted on the claim of Platonists, old and new, that mere ideas can generate real matter, including its totality, the Universe. The erstwhile version of this Platonism need not detain us. It has, however, today a novel appeal dressed as it is in esoteric references to world-lines, zero-point oscillations in the quantum mechanical vacuum, and the like. Because of their scientific glitter, these expressions are apt to be taken for factors that can generate material reality and its totality as well, or the Universe. Unfortunately, credence is given even to that crudest renewal of primitive Platonism which makes some modern scientific cosmologists claim, as has been seen, that by scribbling complicated equations they can literally create universes out of nothing. Only by falling head over heels for that glitter will one find less convincing the starting point riveted in real matter.

Such is a starting point so elementary as to admit but rudimentary demonstrations. Samuel Johnson offered a memorable one when he kicked a stone. A genuine *experimentum crucis* it was, unconvincing only for those unable to see the *crux* of the matter. Although the experiment must have been repeated many times, it is not known to have convinced idealists, Berkelian or other. They will not be moved on hearing that however hard they may think about stones, only their reasoning will be hardened thereby, while they may be reduced to stony silence about real stones. Philosophy, unlike religion, is an area where conversions hardly ever occur and if they do they benefit intellects much more pliable than are leading figures of non-realist philosophies. They all are ready to surrender to all truth except to the truth about matter or external reality.

The truth of an external reality, which exists whether one thinks about it or not, is so elementary as to admit no detailed discourse unless the details become repetitions of the same. Conviction about that truth can, however, be greatly strengthened by looking at some of the consequences that result when the opposite is claimed, let alone when the claim includes a combination of scientific glitter and some apparent philosophical sophistication. Such is the declaration of a prominent scientific cosmologist: "No phenomenon is a phenomenon until it is observed. . . . The universe does not 'exist out there' independently of all acts of observation. It is in some strange sense a participatory Universe."[1] About a Universe which is "participatory" in this idealist sense one should at least note that it cannot call for a proof of its existence without destroying the idea of participation. The latter makes sense only if the thing which is taken part of exists even when the act of participation does not take place. Otherwise one should merely speak of the extrapolation if not morbid protrusion of the self.

The other kind of discourse that should be particularly suspect is the Kantian version of Platonism, which has several self-defeating, and therefore very instructive, aspects to it. These if kept in mind can strengthen conviction about taking real matter for the starting point in reasoning about anything, and certainly about the Universe. One of those aspects is Kant's claim that outer reality cannot be known in itself. He wrapped that claim within his notorious sophistry about the "Ding an sich." It should be enough to recall that the Sage of Königsberg was at a loss for a reply when some of his students asked him for his reasons to speak of something that cannot be known. They merely anticipated Wittgenstein's elementary wisdom although it may have bounced off Kant's idealist hide without making the slightest dent upon it.

It is this unknowability of the "Ding an sich" which explains Kant's caginess about real matter, about its countability, and, last but not least, about the impossibility of an actually realized infinite quantity. Had he known the sleight-of-hand which Halley offered about that quantity, and in a cosmological context at that,[2] Kant might have taken special comfort. He must have been quite uncomfortable with the fact, obviously known to him, that Lambert endorsed the finiteness of the universe with a view on exactly that impossibility.[3] As it turned out, no mathematician of the stature of Lambert was ever impressed by Kant's tactic to cover up ignorance of higher mathematics in artfully cryptic utterances about the infinite. They are like mist when exposed to the stark brilliance of the dictum of Gauss, the prince of mathematicians: "I protest against the use of an infinite magnitude as something completed, which is never allowed in mathematics. The infinite [in mathematics] is a mere *façon de parler*."[4]

The perceived validity of the first antinomy will remain in inverse proportion to the strength of conviction whereby one ties the impossibility of an actually realized infinite quantity to countable real matter. One should indeed be resolved to take a very firm stand along the entire line of the argument or proof, of which that countability is a part, about the reality of the universe. It is most dangerous to weaken the thrust of Kant's use of that antinomy by saying, as did Pamela M. Huby, that "while it is true that Kant was concerned to show the limitations of pure reason, he also used the antinomies as evidence that the Universe was not fully real."[5] Kant's aim was to show that the Universe was fully unreal as far as critical or "pure" reason was concerned. He might have sharply reacted against the ambivalent expression, "not fully real." It was logically as objectionable as to speak of "almost pregnant" women.

The foregoing vacillation about the true thrust of the first antinomy is all the more regrettable because it was preceded by Huby's resolve not to yield on the pivotal point: "What I am concerned with primarily is the application of numbers in counting, or the application of mathematics to the world."[6] She could rightly suggest that potentially infinite numbers, or Cantor's transfinite numbers, are of interest to pure mathematicians, but not to physicists dealing with real matter. This latter point had, of course, been made with particular force by no less a mathematician than Hilbert according to whom such a potential infinity "is nowhere to be found in reality."[7]

As it turned out, physical reality is not the embodiment of three-dimensionality by which Kant had set so great a store and which could trap the unwary into thinking that a finite universe is impossible. There is more than what meets the eye in Olbers' declaration that the universe has to be infinite because Kant said so.[8] It should indeed seem telling that Kant was remembered, first by his German compatriots and then by countless non-German admirers of his, as the champion of the infinite Euclidean universe. Kant's insistence on Aristotelian categories as necessary or a priori forms of understanding proved to be far more effective than his contrived doubts about the universe. One of those categories, as Kant formulated them, was the three-dimensional perception of space. From this it followed that the universe had to be necessarily infinite in a three-dimensional sense. As it happened all too often in the history of philosophy, once more the master's disciples had to spell out an inference which the master, here Kant, would have refused to draw. The inference contradicted Kant's central claim, made in the first antinomy, that since it was undecidable whether the Universe was infinite or not, its notion was unreliable and, as a consequence, its very reality could not be known with rational certainty.

Conviction about the unreliability of Kant's arguments against the universe can be further weakened (and inversely, conviction in the opposite sense further be strengthened) through some reflection on another though very scientific development. Only a few years after Olbers endorsed cosmic infinity by falling back on Kant's authority, Gauss dissented though only in private. Gauss first did so in a letter he wrote on April 9, 1830, to Bessel where his reference to Kant was only indirect: "We must humbly admit that, even if numbers were to be the mere products of our intellect, space must have, independently of our thinking, a reality whose laws we cannot easily prescribe a priori."[9] The context of Gauss' remark related to his efforts to find an appreciable departure from 180 degrees, which is the sum of angles in a flat triangle, from the sum of angles in a triangle on a curved surface. This was also the background of his second observation on the subject, with Kant now being explicitly named. In his letter of March 6, 1832, to Wolfgang (Farkas) Bolyai, Gauss referred to that experimental difficulty and noted that far greater was the difficulty of giving a proper definition of a plane: "This difficulty is altogether different from the difficulty of deciding between Σ and *S* [the two respective sums of angles]. Precisely in the difficulty of deciding on an a priori basis between Σ and *S* lies the clearest proof that Kant was wrong in asserting that space is just a form [category] of our perception."[10] Kant was not only mentioned but also held up to slight ridicule in Gauss' letter of Nov. 1, 1844, to the astronomer, H. Schumacher. There he took the view that Kant's distinction "between analytic and synthetic propositions is one of those things that either run out on triviality or are false."[11]

While these statements of Gauss became available in print only in the closing decades of the 19th century, Neo-Kantians did not wish to be convinced about some basic fault-lines in the Kantian edifice of knowledge. They paid equally scant attention when in the early 1870s a specific proposition was made about the application of non-Euclidean geometries to scientific cosmology.[12] The attitude of Neo-Kantians may suggest that Kant himself would have acted in a similar way, although only by putting at grave risk his well-nurtured claim about being a Newtonian, that is, scientific metaphysician. Half a century later Neo-Kantians found just as unconvincing the cosmological thrust of Einstein's general relativity and Einstein's emphasis on the non-Kantian character of the free creations of the intellect needed in creative physical science.[13] Perhaps it was best to keep under cover the fact that the application of non-Euclidean geometries in cosmology, which took full form in general relativity, restored scientific dignity (and therefore intellectual dignity in this age of science) to the notion of the universe.

The restoration of the notion (and reality) of the universe into scientific dignity is not the same as assuring philosophical respectability to a proof of the reality of the universe. Science, or rather the exact science of physics, is not about reality as such but only about the quantitative aspects of things already in existence. Nothing poses so much threat to achieving philosophical conviction about anything as the concession that the scientific method is the key to reality. In the process science is perverted into a gnosis, while philosophy is turned into a ghost to be legitimized by science and be allowed to parade as a "scientific" philosophy. In terms of a perhaps more convincing, because more popular, analogy, the outcome is to cast philosophy in the role of a baseball team that is assured the right to start its runners from second base, while the ball is never hit and science is hired as an umpire to complete the parody. The analogy could be set in terms of a soccer club that is allowed to start from the end of the penalty box though never given a ball to kick. It became the glory of a prominent cricket captain to offer, however unwittingly, perennial food for thought to "critical" philosophers with his remark: "All of sport is about the initial movement and if you interfere with that you are in trouble."[14] The entire history of modern philosophy is a proof of that trouble, though instructive only to those fond of registering reality right at the outset of any philosophical discourse.

The registering of reality is a primary act that cannot be proven in terms of anything else. It is irreducible to anything else on the level of knowledge which is communicable and communicated. Reality no more can be proved in terms of something else than truth in terms of error, unless the error is taken for truth. The probable cannot serve as the basis of what is certain, or else one loses all certainty. At times even those most suspicious of all certainties can spot this elementary truth though not without trying to eschew its significance. A case in point is David Hume's acknowledgment: "Probability is founded on the presumption of a resemblance betwixt those objects of which we have had experience, and those of which we have had none; and, therefore, it is impossible this presumption can arise from probability."[15]

This, however, means that the certainty of real matter is a primary datum which cannot be derived from probable occurrences of it, let alone from their mere ideas. Enormous should seem the instructiveness of this truth for strengthening conviction in the proof of the existence of the universe which begins with real, countable matter. If one admits that probability cannot have the first word with respect to reality, one will not be overawed by the smudging of the Universe into ensembles of probabilities, as all too often happens in modern scientific cosmologies. It will then easily be perceived that such a procedure, glitteringly mathematical as it may be, has for one of its aims the glossing over of the radical difference between infinity and finiteness. But if the material content of an ensemble is left undefined, the continued postponement of taking it for a definite reality puts one on a road where regress to infinity is the tacit policy.

Temptation to choose that road must be resisted from the outset if conviction in the proof, as developed in the previous chapter, is to be achieved. Nurturing that conviction will be nipped in the bud, if the universe is to be reached from a starting point other than concrete countable matter. Space has been taken, time and again, for such a starting point and with patently disastrous consequences. Suffice to think of Lord Kelvin's declaration before a festive audience in Philadelphia that "*finitude* is incomprehensible, the infinite in the universe *is* comprehensible."[16] His proof rested on invoking the inability of imagination to set a limit to space. To heighten the irony, he failed to see that in claiming this he blinded himself to the truth of cosmic finitude. As he listed various components of that truth, he turned them inside out.

It was certainly easy to imagine with Lord Kelvin that one could "travel one, ten, or a thousand miles, or even to California" and find it stultifying that one would thereby "find it [the universe] come to an end." It was equally easy to proceed in imagination to ever larger numbers, "from one to ten, and ten times ten and then to a thousand . . . and then . . . to a thousand million and a million million." But this did not mean that one could thereby count infinity as a number. That Lord Kelvin failed to see this elementary point may have its explanation in his focusing on space to which imagination could not set bounds. He lost out by the same stroke on a real material universe. The irony was all the more complete because he also raised, in the same breath, the question in which matter and space stood in the right sequence, though not implying the primary importance of the former: "Can you suppose an end of matter or an end of space?"[17] It escaped Lord Kelvin that matter, insofar as it was countable, had to set a limit to space whatever it was.

Taking space for a starting point in thinking about the universe could mislead minds philosophically far more sophisticated than Lord Kelvin's. Too much attention given to space can invite too early in the discussion Cantor's transfinite numbers, and intuition to boot. Thus even when it is argued, and correctly so, that all that Cantor did was to introduce techniques that allow the unambiguous handling of "a certain kind of *potential* infinite," the conclusion that the world must be finite "in space and time" may turn into a boomerang against the reality of the universe.[18]

By making the universe a function of matter one can avoid doing a signal injustice to general relativity. Such an injustice is in the making if one starts with a space (infinite or not) to be filled with matter. A basic achievement of general relativity was to demote space from a pedestal where it posed as a reified idea. Whatever can be said by philosophers about space, in relativistic cosmology, as first conceived by Einstein, space can be no more than a network of permissible paths of motion, a mere mathematical abstraction to which nothing real corresponds in the physical world.

This genuinely Einsteinian way of putting space in its place, through general relativity, failed to set an effective barrier against the temptation of taking space for a container and for something much more. Before long, modern physicists, who had just got rid of the ether, began to reify space as if it could exist apart from matter. In fact, all too often they went on to suggest that it is curved space that generates matter and its totality, the universe. While such a suggestion could perhaps be explained by assuming that space was taken for gravitational field, the latter could only be had if there was already matter on hand. Otherwise the claim amounted to putting the cart before the horse, splendid as both could be in their abstract mathematical formalisms.

The reification of space by prominent students of general relativity led also to another cosmological misconception, which ought to be kept in focus if our proof about the reality of the universe is not to be threatened in its convincing character. The misconception relates to the possibility of imagining models of space which are, unlike the Riemannian four-dimensional spherical space, open-ended. Cylindrical and hyperbolic spaces are cases in point. The cosmologist is at liberty to fill such spaces with an infinite amount of matter if he has to avoid only the problem arising from the gravitational paradox in a Euclidean space or in a spherical Riemannian space. But the problem is avoided only because in a cylindrical and a hyperbolic space the inverse square law of gravitation cannot form an unambiguously central field of force. It is, however, invariably overlooked that this does not dispose of the problem of an actually realized infinite quantity.[19] Open-ended as can be a four-dimensional manifold into which matter is cast, the totality of matter it can accommodate cannot be infinite as long as that matter is countable.

This finiteness, or limitedness, is implied also in the specific character of any of the non-Euclidean geometries. They stand for a limitation to one specific net of world-lines or permissible paths of motion. The impossibility of an actually realized infinite quantity will be concretely on hand with concretely existing matter, which must have countable properties. It is well to recall that this embedment of quantities in real sensory matter loomed overriding in Aristotle's discussion of the impossibility of an actually realized infinite quantity.[20] This may further enhance the convincingness of the realist approach which is a hallmark of the philosophical road Aristotle was the first to chart. There is therefore more to the application of that impossibility to the material universe than the comparatively trivial question whether the universe is finite or infinite, that is, bounded or unbounded.

The fact, as was argued earlier, that some modern cosmological models can accommodate an infinite amount of matter without giving rise to the gravitational paradox does not weaken this remark. Those models merely bring out a fact which could not be easily perceived about three-dimensional infinity prior to the advent of non-Euclidean geometries. Since three-dimensionality corresponds to ordinary perception, it could readily be taken for a naturally existing structure, a facet hardly to be ascribed to other structures. The formulation and cosmological application of other geometries—spherical and hyperbolical—should have made it clear that, mathematically speaking, Euclidean space was no less specific than the spaces corresponding to those other geometries which look indeed very specific already at first sight.

This means that each of the infinite space-time manifolds used in modern scientific cosmology is very finite in its being a very specific manifold. Indeed that specificity can in each case be represented by a specific number standing for the curvature specific to that manifold. To ignore this is to ignore a deeper significance of Eddington's remark: "Infinity is the mischief and science must have nothing to do with it."[21] All apparent infinities that are useful for science are subtly confined finitudes.

Subtlety is a category of philosophy. It is therefore most unlikely that this answer to the question of how it can be known that there is a Universe of things, will be seriously considered, let alone be widely accepted in scientific circles, whatever the fairly scientific garb of the answer. In those circles, for now almost four centuries, scientific answers have been given to questions that are essentially philosophical. The scientific approach can go to an impressive length in our subject. It can be most profitably considered why physical interactions are restricted to specific patterns, why matter is limited to a very specific structure as evidenced, for instance, by the selection rules that apply throughout the quantum mechanical domain. The distribution of matter in space is another form of limitation. Wherever found, physical reality looks as if it were caught in a mesh of limitations which may suggest the presence of an overall limitation. Such a limitation provides oneness to the multitude.

To prove on this quasi-scientific ground the reality of an absolute totality and secure to it a convincing character should seem important in an age in which philosophical thinking is uncertain of itself unless confident of its being approved from the scientific side. The proof will appear convincing in the measure in which it is recognized that science cannot raise objections to the idea of a strict totality insofar as it intends to count it. The proof will further gain in convincing character by considering that science, as long as it has to operate with integers, cannot raise objection to the idea of that totality as being a totality consistent throughout.

Convincing character can also accrue to the proof from a brief look at the recently much glorified notion of chaos. It should be realized at the outset that an absolute chaos remains unthinkable. All statistical ensembles that are useful for physics rest on invariable parameters, including the type of randomness ascribed to each ensemble. The idea of chance as used in science does not become clear by the physicist's proverbial refusal to take a close look at what he means by it. Only a very few physicists are ready to admit that, as was noted in the preceding chapter, they use the terms probability "in a far from rigorous fashion."[22] This is even more true when they resort to the words, "chance," "randomness," and "chaos." Were this not the case, there would have been no reason for the Nobel-laureate, Philip W. Anderson, to voice his indignation about the trend of reifying probability and randomness. He deplored the fashion of speaking of "chaos," instead of "deterministic chaos," though he failed to notice that science popularizers merely followed the lead of prominent scientists.[23]

All these details on the quantitatively specific and specifically quantitative nature of all material things will serve as convincing references to reality only when one admits that there ought to be really existing things before there can be really quantitative properties. But this is precisely the condition which cannot be reconciled with the anti-ontological presuppositions of the Copenhagen interpretation of quantum mechanics.[24] There one is invited to blur the difference between being and non-being with that logical result which has recently received its ultimate form in the idea of quantum creation. The lame disclaimers of some of its proponents that quantum creation is not strictly meant to be a creation out of nothing,[25] should appear revolting enough. In fact they are more revolting than the full consistency that inspired an illustration in a recent book on the origin of the universe by three Chinese astrophysicists, one of them Fang Li Zhi, well known in the West for his political stance. The diagram is a glorification of "scientific" creation out of nothing, in the form of a flag with the word NOTHING on it.[26] That diagram can at best globalize an intellectual indigestion about which Robert Louis Stevenson warned a century ago with an eye on those who claimed to have achieved a full scientific account of the origin of the planetary system: "They swallow the universe like a pill."[27]

For a long time now a secularized West has effectively unlearned respect for the marvel of being, or ontological reality. Unable to appreciate that marvel, the Western World had to seek a substitute for it in undue admiration for the metric. The distinction between the metric and the non-metric, in which Eddington rightly saw the difference between the wide humanistic domain and the narrowly scientific one,[28] has now turned into a fashionable denial of the objectivity of the former. A culture imbued with that denial demands "color-blindness" in a far deeper sense than the one acknowledged by Eddington in writing that "there is no colour in the physical world. Everything we [scientists] assert can be verified by a colour-blind person."[29] The idea of a universe produced by quantum creation is the ultimate form of the intellectual tyranny with which a so-called scientific thinking imposes total blindess about ontology.

Liberation from that blindness is the basic condition, not scientific but strictly philosophical, for convincingly using considerations about universals, about material entities with countable properties, about the universal role of integers in the material realm, and about the impossibility of an actually realized infinite number—as so many supports of the argument that there is a strict totality of things. The question about the strict totality of things is philosophical though not in disparagement of the science of quantities. Not science but scientists (and philosophers swayed by their all too often inept philosophizing) formed the target of Gilson's remark that "those who do not wish to hear the answers of metaphysics should not pose questions to it."[30]

When it comes to the universe, Burtt's remark becomes especially true that "the only way to avoid becoming a metaphysician is to say nothing."[31] Much less can one, even if he is a scientific cosmologist, say "everything" or universe if he refuses to be a metaphysician. It does more harm than good if the metaphysics in question is the one that begins with the mind, instead of with the objectively material. Those who started with the mind, never reached matter, a point still to be learned by scientific cosmologists through a careful study of the history of Western philosophy. But if there is no physical realm, one tries in vain to go beyond physics in that sense which is metaphysical. So much in a way of comment about Burtt's other remark, appropriate both to philosophers and scientific cosmologists, that "an adequate cosmology will only begin to be written when an adequate philosophy of the mind has appeared."[32]

What is really needed is a recovery of the sense of the real from the clutches of rank idealism and blissful endorsements of Platonism as if equations and co-ordinate systems were the foundation of reality and therefore the creators of the universe itself. Until that happens, a prominent cosmologist may but momentarily awaken from the slumber of his philosophical idealism and register his astonishment over material reality by exclaiming: "The Universe flies!" That slumber must be totally dissipated if a step is to be taken from wondering about reality to the reality of the Universe itself. For if in the courtroom of science, cosmologists will testify to the reality of the Universe, it may happen that even philosophers and theologians will admit that what flies is not merely anything but the "everything," or the Universe itself.

This and similar perspectives do not become less worthy of consideration just because they were first formulated with an eye on the reality of the One who can create only in such a rational way as to assure to His products that unity which is the only safeguard of understanding. Every word is a proof, a fact recognized already by Plato. The one and the many, so one reads in *Philebus*, "run about together, in and out of every word which is uttered."[33] In a more technical parlance, each word is a universal, a means to grasp much more than the single item actually seen. This surplus is also a means to see the one in the many. Knowledge is therefore inferential even when no distinct steps of inference are performed. The very fate of knowledge is indeed at stake when knowledge is not trusted whenever its object of inference is the totality of things or the Universe.

Knowledge is honored in its very fundamentals when one upholds the validity of inference to the most universal of all universals which is the universe of things. Once there, knowledge is crowned when it implies the inference to the reality of the One which is such because it is its very Being to exist. As such He has that absolute power over being which is needed to produce other beings and their totality, the universe. The universe itself is raised to the highest conceivable pedestal—and a very safe one at that where no intellectual vertigo threatens the cosmologist—if one says, with John Henry Newman, that the idea of the Universe is so great that only the idea of its Maker is greater.[34] Had Newman been an idealist, he would have thereby invited a mental vertigo. What he really meant echoed a long-standing conviction, voiced among others by Aquinas, that even the Creator could not have created anything greater than the Universe. Such has to be the case if the Universe, or the converging of all, is a reflection of the coherence which God has to be. The science of cosmology is unfolding magnificent vistas about the coherence of everything under its purview, while it has to take it for granted that there is such a totality of things which deserves to be called the Universe. This ultimate physical entity will loom convincingly at the end of one's mental journey as long as one holds fast to the right starting point.

[1] A statement of J. A. Wheeler, quoted in the editorial to the May 1986 issue of *American Journal of Physics*, by its editor, J. S. Rigden (vol. 54, p. 397).

[2] In the first of the two papers Halley read in 1720 before the Royal Society, Halley acknowledged that the view according to which the number of stars "is not only indefinite, but actually more than any finite Number" may seem to be "absurd *in terminis*, all Number being composed of Units, and not two Points or Centers being at a distance more than finite." In the second paper Halley characterized the infinity of a given number as a "metaphysical paradox." He countered that if that view were true, there could be no infinite duration, or eternity. As an atheist, Halley wanted to turn the tables on believers while forgetting that eternity was different from infinite duration. Both papers of Halley are reprinted in my *The Paradox of Olbers' Paradox* (New York: Herder & Herder, 1969), pp. 249-52.

[3] See its English translation, *Cosmological Letters on the Arrangement of the World-Edifice*, tr. with an introduction and notes by S. L. Jaki (New York: Science History Publications, 1976), pp. 47, 125 and 216. The book, it may be noted, is, within its scientific context, the first contradiction-free treatment of the totality of gravitationally interacting things. In 1761 it was not yet imperative to worry about the dissipation of gravitational energy through motion in interstellar space.

[4] Gauss' letter of July 12, 1831, to Schumacher. See *Briefwechsel zwischen C. F. Gauss und H. C. Schumacher* (Altona: Gustav Esch, 1860), vol. 1, p. 269.

[5] P. M. Huby, "Cosmology and Infinity," *Philosophy* 48 (1973), p. 186.

[6] Ibid., p. 187.

[7] D. Hilbert, "On the Infinite," in *Philosophy of Mathematics*, ed. P. Benacerraf and H. Putnam (Englewood Cliffs, N. J.: Prentice Hall, 1964), p. 151.

[8] Tellingly, Olbers quoted from Kant's *Allgemeine Naturgeschichte,* a scientifically as well as philosophically amateurish work. See my *The Paradox of Olbers' Paradox*, p. 257.

[9] *Carl Friedrich Gauss — Friedrich Wilhelm Bessel. Briefwechsel* (Hildesheim: Georg Olms, 1975), p. 497.

[10] *Briefwechsel zwischen Carl Friedrich Gauss und Wolfgang Bolyai,* ed. F. Schmidt and P. Stäckel (Leipzig: B. G. Teubner, 1899), p. 112.

[11] *Gauss. Gesammelte Werke* (Göttingen: K. Gesellschaft der Wissenschaften, 1870-1933), vol. 12, p. 63.

[12] Its author was Zöllner, discussed in ch. 1. About the same time, E. Schering, editor of Gauss' scientific writings, published two short essays on gravitation and non-Euclidean geometry in the *Nachrichten* of the Academy of Sciences and the University of Göttingen (July 13, 1870, pp.

311-21 and February 26, 1873, pp. 149-59), but they did not deal with the cosmological problem.

[13] Einstein had in mind the free creativity of the mind, as revealed in his general theory of relativity, in noting that "nothing can be said *a priori* concerning the manner in which the concepts are to be formed and connected. . . . Their fixation will never be final. It will have validity only for a special field of application (i. e. there are no final categories in the sense of Kant)." *Ideas and Opinions of Albert Einstein*, p. 292. Duhem had by then spoken in his *La théorie physique* (1906) of the role which is played in physics by the free creations of the mind.

[14] Bob Simpson, the captain of the Australian cricket team, following his team's devastating triumph over England in August 1989.

[15] D. Hume, *A Treatise of Human Nature*, Book I, Part III. Sec. vi.

[16] W. Thomson, "The Wave Theory of Light," in *Popular Lectures and Addresses* (London: Macmillan, 1889-94), vol. I, p. 314.

[17] Ibid.

[18] As is clear from P. M. Huby's essay, "Kant or Cantor? That the Universe, if Real, Must Be Finite in Both Space and Time," *Philosophy* 46 (1971), pp. 121-32, and W. H. Newton-Smith's reply, "Armchair Cosmology," ibid., 47 (1972), pp. 64-70.

[19] A point overlooked, for instance, by S. Weinberg, *Gravitation and Cosmology: Principles and Applications of the General Theory of Relativity* (New York: John Wiley, 1972).

[20] See *Physics*, III, 5-6 (204a-206b).

[21] A. S. Eddington, *New Pathways in Science* (Cambridge: University Press, 1934), p. 217.

[22] P. G. Bergmann, *The Nature of Time* (Ithaca, N.Y.: Cornell University Press, 1967), p. 109.

[23] P. W. Anderson, "On the Nature of Physical Laws," *Physics Today*, Dec. 1990, p. 9. One of those great scientists was Einstein. He carelessly equated "chaos" with "statistical laws" as he wrote to J. Franck: "I can imagine that God created a world without any laws: a chaos in short. But the notion that statistical laws are final and that God draws a lot, is highly unsympathetic to me." Quoted in Y. Elkana, "The Myth of Simplicity," in *Albert Einstein. Historical and Cultural Perspectives. The Centennial Symposium in Jerusalem*, G. Holton and Y. Elkana (eds.) (Princeton, N. J.: Princeton University Press, 1982), p. 239.

[24] The consistency with which Bohr used his principle of complementarity to avoid questions of ontology is taken, and rightly so, for the pivotal point of his philosophy in A. Hooker's vast essay, "The Nature

of Quantum Mechanical Reality," in R. Colodny (ed.), *Paradigms and Paradoxes: The Philosophical Challenge of the Quantum Domain* (Pittsburgh: University of Pittsburgh Press, 1972), pp. 67-302, see especially p. 208. I argued the same point in ch. 13, "The Horns of Complementarity," of my Gifford Lectures, *The Road of Science and the Ways to God* (Chicago: University of Chicago Press, 1978), and more recently in my essay, "Determinism and Reality," see note 12 in ch. 3 above.

[25] Thus, for instance, C. J. Isham, "Creation of the Universe as a Quantum Process," in *Physics, Philosophy, and Theology: A Common Quest for Understanding*, R. J. Russell, W. R. Stoeger and G. Coyne (eds.), (Vatican Observatory, 1988), who defeats his disclaimer by claiming "theological implications" for that "creation" (p. 404). The latter can be called creation only with a brazen disregard of what is at stake for theology in keeping uncontaminated the idea of creation out of nothing.

[26] Fang Li Zhi and Li Shu Xian, *Creation of the Universe*, tr. T. Kiang (Singapore: World Scientific, 1989), p. 147. It is stated there that prior to 1982 the relation of non-being to being was considered a purely metaphysical one, but the Symposium held in that year in Cambridge on the very early universe provided "several key answers given by physics" to the question of "being from non-being." Of course, what ultimately all those answers offered was a spatialization of 'nothing' as something that exists 'outside' the universe that has no boundaries. It is rather sad that physicists are proud of such invasions of theirs into metaphysics which they conquer by dissipating it into games with geometry.

[27] "Crabbed Age and Youth" (1878), in *Essays by Robert Louis Stevenson*, with an Introduction by W. L. Phelps (New York: Charles Scribner's Sons, 1918), p. 136.

[28] "The cleavage between the scientific and extra-scientific domain of experience is, I believe, not a cleavage between the concrete and transcendental, but between the metrical and non-metrical," wrote Eddington in his *The Nature of the Physical World* (Cambridge: University Press, 1928), p. 275.

[29] *New Pathways in Science* (Cambridge: University Press, 1935), p. 11.

[30] The closing phrase of Gilson's contribution, "En marge d'un texte," to the volume, *Louis de Broglie: Physicien et Penseur* (Paris: Albin Michel, 1953), p. 158.

[31] E. A. Burtt, *The Metaphysical Foundations of Modern Physical Science* (2d rev. ed., 1932; Garden City, N. Y.: Doubleday, n. d.), p. 227.

[32] Ibid., p. 324.

[33] *Philebus*, 15d.

[34] J. H. Newman, *The Idea of a University* (8th ed.: London: Longmans, Green and Co., 1888), p. 462.

Index of Names

Index of Subjects

(continued from p. ii)

By the same author

The Savior of Science
(Wethersfield Institute Lectures, 1987)

Miracles and Physics

God and the Cosmologists
(Farmington Institute Lectures, Oxford, 1988)

The Only Chaos and Other Essays

The Purpose of It All
(Farmington Institute Lectures, Oxford, 1989)

Catholic Essays

Cosmos in Transition: Studies in the History of Cosmology

Olbers Studies

Scientist and Catholic: Pierre Duhem

Reluctant Heroine: The Life and Work of Hélène Duhem

Universe and Creed

Genesis 1 through the Ages

* * *

Translations with introduction and notes:

The Ash Wednesday Supper (Giordano Bruno)

Cosmological Letters on the Arrangement of the World Edifice (J.-H. Lambert)

Universal Natural History and Theory of the Heavens (I. Kant)

Note on the Author

Stanley L. Jaki, a Hungarian-born Catholic priest of the Benedictine Order, is Distinguished University Professor at Seton Hall University, South Orange, New Jersey. With doctorates in theology and physics, he has for the past thirty years specialized in the history and philosophy of science. The author of over thirty books and nearly a hundred articles, he served as Gifford Lecturer at the University of Edinburgh and as Fremantle Lecturer at Balliol College, Oxford. He has lectured at major universities in the United States, Europe, and Australia. He is honorary member of the Pontifical Academy of Sciences, *membre correspondant* of the Académie Nationale des Sciences, Belles-Lettres et Arts of Bordeaux, and the recipient of the Lecomte du Nouy Prize for 1970 and of the Templeton Prize for 1987.